HOW PATHOGENIC VIRUSES *THINK*

Making Sense of Virology

SECOND EDITION

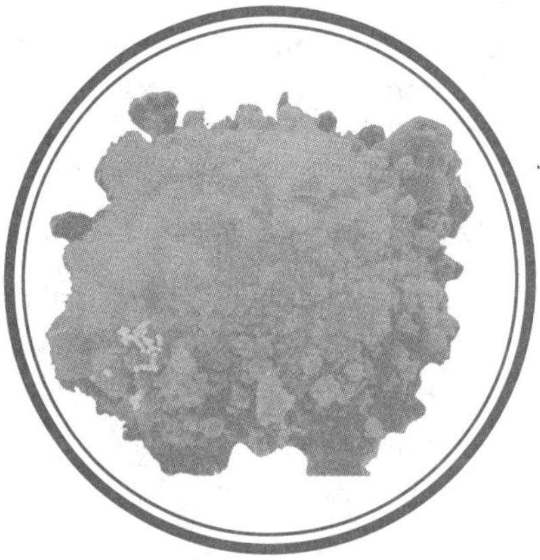

Lauren Sompayrac, PhD

Retired Professor
Department of Molecular, Cellular, and Developmental Biology
University of Colorado
Boulder, Colorado

JONES & BARTLETT
LEARNING

World Headquarters
Jones & Bartlett Learning
5 Wall Street
Burlington, MA 01803
978-443-5000
info@jblearning.com
www.jblearning.com

Jones & Bartlett Learning books and products are available through most bookstores and online booksellers. To contact Jones & Bartlett Learning directly, call 800-832-0034, fax 978-443-8000, or visit our website, www.jblearning.com.

Substantial discounts on bulk quantities of Jones & Bartlett Learning publications are available to corporations, professional associations, and other qualified organizations. For details and specific discount information, contact the special sales department at Jones & Bartlett Learning via the above contact information or send an email to specialsales@jblearning.com.

Production Credits
Executive Publisher: Christopher Davis
Managing Editor: Kathy Richardson
Editorial Assistant: Marisa LaFleur
Production Assistant: Sarah Burke
Marketing Manager: Rebecca Rockel
Manufacturing and Inventory Control Supervisor: Amy Bacus
Composition: Jason Miranda, Spoke & Wheel
Cover Design: Scott Moden
Cover Image: Courtesy of Jim Adams
Printing and Binding: Edwards Brothers Malloy
Cover Printing: Edwards Brothers Malloy

Library of Congress Cataloging-in-Publication Data
Sompayrac, Lauren.
How pathogenic viruses think : making sense of virology / Lauren Sompayrac.—2nd ed.
p. ; cm.
Rev. ed. of: How pathogenic viruses work / Lauren Sompayrac. c2002.
Includes index.
ISBN 978-1-4496-4579-3—ISBN 1-4496-4579-8
I. Sompayrac, Lauren. How pathogentic viruses work. II. Title.
[DNLM: 1. Viruses. 2. Virus Diseases. QW 160]

616.9'101—dc23

2012003777

6048

Printed in the United States of America
16 15 14 13 12 10 9 8 7 6 5 4 3 2 1

DEDICATION

I dedicate this book to my outstanding editor, Chris Davis, who shares my belief that although science is serious, it should also be fun. Thank you, Chris!

Contents

Part II: The Bug Parade *19*

Viruses We Inhale

Viruses We Eat

Viruses We Get From Mom

Chapter 10 Hepatitis B: A Decoy Virus *73*

Chapter 11 Hepatitis C Virus: An Escape Artist *79*

Chapter 12 HTLV-I: A Tribal Virus *85*

Viruses We Get By Intimate Physical Contact

Part III: Beyond the Bug Parade *121*

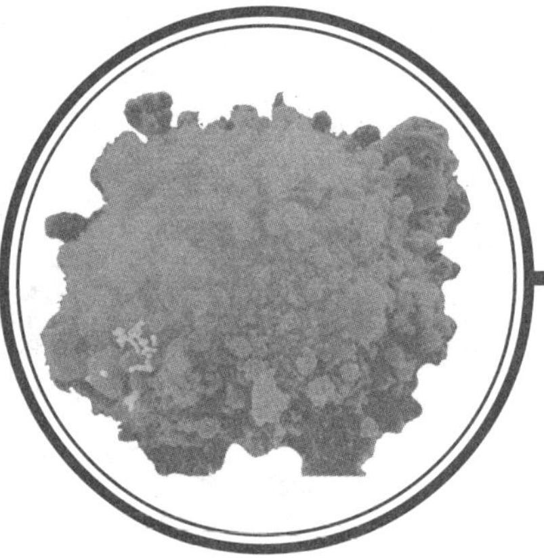

Acknowledgments

I especially want to thank my good friend, Bob Mehler, who read the entire first edition, and who offered excellent advice and suggestions. I also wish to thank the following people who offered critical comments: Charles Bangham, Jim Cook, Andreas Dotzauer, Bin He, Thomas Hope, John Kash, Karla Kirkegaard, Mari Manchester, Jack Routes, and Aleem Siddiqui. Thanks also to my lovely wife, Vicki Sompayrac, whose wise suggestions helped make this book more readable, and whose editing was invaluable in preparing the final manuscript.

I would like to express my gratitude to Richard Feldmann, Shpilke Rozenblatt, and Nick Wrigley for contributing some of the figures in this book. Their beautiful pictures truly are worth thousands of my words. And finally, I would like to thank Jim Adams, who was the illustrator for the first edition, and whose wonderful drawings, including the cover image, continue to grace this second edition.

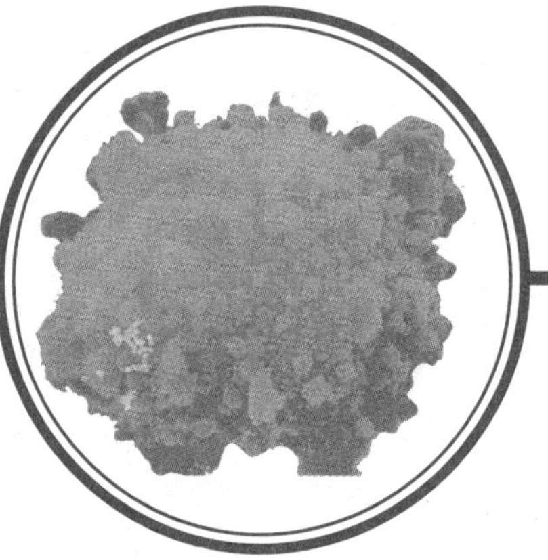

Introduction

When I first studied virology, there were only a few viruses about which much was known—so it was relatively easy to get one's mind around this subject. Things are very different now. Today, most popular virology textbooks boast more than 1,000 pages, crammed full with details about almost any virus in which one might have an interest. This begs the question: How can a virology student possibly deal with all this information?

Part of the answer is that to make sense of virology, you need an "organizing principle." In *How Pathogenic Viruses Think*, I introduce such a principle—a paradigm which you can use to analyze any virus. Understanding viruses is similar to understanding people from different countries. All humans have a common set of problems which we must solve to survive. However, because our geographies, natural resources, and histories vary, the ways people from different countries solve these problems will be different. Likewise, viruses must solve a small number of common problems if they are to survive. By focusing on these problems, and by asking how a given virus solves them, we can make sense of the wealth of information about viruses and appreciate the clever solutions viruses have come up with.

So the first difficulty we face in making sense of virology is to discover an organizing principle we can use to deal with all the data which has been amassed. But there is a second problem. Most textbooks are written from the viewpoint of an "outside observer"—the virologist who performs experiments and collects information about different viruses. The result usually is a book filled with facts, lists, and experimental techniques. These texts will tell you all about what a virus does, but they rarely tell you WHY the virus does it that way. And if you don't know why something happens, it becomes a matter of memorizing, not learning. *How Pathogenic Viruses Think* is different, because our goal is to "fathom the mind of the virus"—to examine the virus-host interaction from the virus' perspective, always asking why the virus has chosen to do things "his way."

In the first chapter, I will introduce a paradigm which you can use to cut through all the details, and focus on what's important. Then, in subsequent chapters, we will use this paradigm to examine the lifestyles of 12 common pathogenic viruses. And to make sure that we are looking at things from the virus' perspective, we will let each virus "tell his own story." Viruses have slightly different "personalities" because of the ways they have evolved to solve their common problems, and these personalities will be revealed as the "Interviewer" asks each virus to divulge his secrets. I think you will discover that when you use this paradigm to study virology from the virus' point of view, the subject will make sense, the facts will fall naturally into place, and the diseases which result from a viral infection will be understandable as a consequence of the way the virus has chosen to solve its problems.

Although an innovative professor might use this book as the text for a one-term virology course or for a section in a medical microbiology course, *How Pathogenic Viruses Think* really isn't intended as a textbook. It is meant to be the book students read to help them make sense of all the information in the big textbook. However, no matter how your professor may choose to use this book, you should keep one thing in mind: I didn't write this book for your professor. This book's for you!

Lauren Sompayrac, PhD

Part I

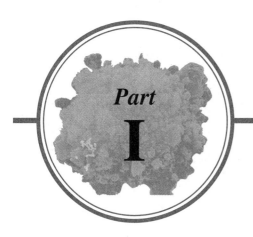

Fathoming the Mind of a Virus

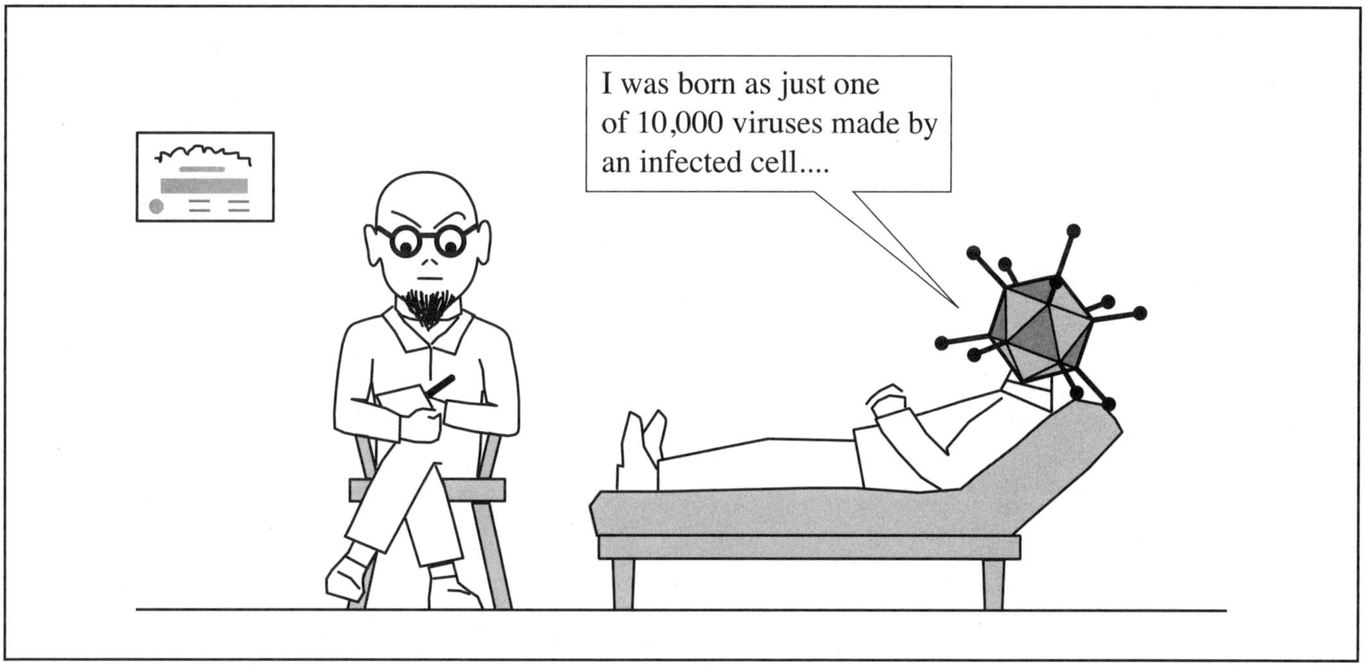

We will begin our quest to "fathom the mind of a virus" with three important introductory chapters. In the first, I'll give a little background on what viruses are, and where they may have come from. Then I will introduce a paradigm which can be used to analyze any virus whose mind you may wish to fathom — an organizing principle which can help make viruses (and virology) more "accessible."

In Chapters 2 and 3, I will discuss the potent, multi-layered defenses humans have evolved to defend themselves against viral attacks. These defenses, and a virus' response to them, play a critical role in shaping a virus' lifestyle. Consequently, if you want to know how viruses "think," you will need to understand the threats they face from host defenses. As any virus will tell you, "It's a dangerous world in there!"

Chapter 1 The Organizing Principle

BACKGROUND

Viruses are pieces of RNA or DNA enclosed in a protective coat(s). These simple organisms are parasites which have evolved to reproduce inside, and survive outside, the cells they infect. What makes viruses so amazing is that they can do so much with so little—and that they do it so elegantly. For example, in terms of genetic information, hepatitis B virus is the smallest known human virus (i.e., virus that infects humans) with only four genes. Yet despite its dearth of genetic information, hepatitis B virus is one of the world's most deadly pathogens: Over a million humans die each year from hepatitis B–associated liver disease.

Contrary to what we see on TV, viruses are not "gifts" to humans from extraterrestrials. In fact, the viruses which now plague humans almost certainly arose from within the cells that make up plants, humans, birds, and animals. I say, "almost certainly," because nobody really knows for sure how viruses evolved—there are no fossil-like records which can be consulted. My view is that viruses arose because Mother Nature was in a hurry. Here's what I mean.

During evolution, mutations occur in the genetic code of an organism, and the "fittest" of the resulting mutants is selected for survival. If these changes to the genetic code had been made only one or a few letters at a time, the evolutionary process would have taken forever. So to speed things up, Mother Nature decided to allow whole "phrases" of genetic information to be transferred from one chromosomal location to another. The use of these "jumping genes" made it possible for complete genetic units to be combined to produce proteins that were multifunctional, or to allow parts of different genes to be spliced together to create proteins that could perform brand new functions. Indeed, it is estimated that over 30% of all human genes bear traces of such "transposition" events.

This gene shuffling greatly accelerated the evolutionary process, but it also provided a mechanism for creating viruses. During transposition, genetic information floats free in the nucleus of a cell, unattached to any chromosome. And since viruses are nothing more than snippets of RNA or DNA enclosed in a protective coat, some of this mobile genetic information could have been used by "wannabe" viruses to construct their **genomes** (defined as the sum total of a virus' genetic information). Of course, any mechanism that resulted in genetic information (either RNA or DNA) being "loose" within a cell could have given viruses their start. However, because viral reproduction requires the functions of multiple viral genes, transposition also made it possible for these "useful" genes to be moved adjacent to each other on a chromosome, so that they could be picked up by a wannabe virus as a group. Based on these considerations, I think it can reasonably be argued that the jumping genes which sped up the evolutionary process provided the raw materials needed for the construction of viruses, and that viruses arose as an unavoidable consequence of more rapid genetic evolution.

If you ask the average person how many different viruses cause disease in humans, he could probably name fewer than a dozen. In actuality, there are more than 50 different viruses which can cause human disease. Because there are so many of them, we can conclude that the evolution of pathogenic (disease-causing) viruses was not a rare event that happened only a few times long ago. In fact, many human viruses probably evolved during the last five to ten thousand years—a mere blink of the eye on the evolutionary scale. And viruses are still evolving today.

One thing to keep in mind is that viruses are very selfish. There are bacteria (e.g., bacteria which colonize the intestine) which are essential for human survival. In contrast, no virus is known to contribute in any positive way to human health. Viruses are in it for themselves.

FOUR PROBLEMS EVERY VIRUS MUST SOLVE

Viruses are real "individuals." Even viruses that belong to the same family (e.g., the herpes virus family) can have dramatically different lifestyles. Fortunately, there is an organizing principle which can help us make sense of all this diversity. That principle is based on the fact that there are four basic problems which every virus must solve. Although it is impossible to trace with certainty the exact events that led to the "birth" of any particular virus, it is clear that the viruses we know today were selected from a much larger crowd of wannabe viruses. Those viruses that survived this selection are the ones which "learned" to solve these four problems.

Infection of Host Cells

The first problem every virus must solve is how to access and infect its target cells. This "decision" is critical, in part because the entry route that a virus chooses will determine which host defenses it must overcome in order to be successful. There are four major pathways for infection which viruses have evolved to use: Viruses can be inhaled, viruses can be ingested, viruses can be passed from mother to child, and viruses can be acquired by intimate physical contact. Each pathway has its own unique set of host defense mechanisms.

Once a virus enters its new host, it must locate cells in which it can reproduce efficiently, and the choice of entry point will determine which cells are available for infection. As a rule, viruses are pretty picky about which cells they infect—not just any cell will do. To be an appropriate target for viral infection, a cell must have receptors on its surface to which the virus can attach. In addition, the biosynthetic machinery within the cell must be compatible with the reproductive strategy used by the virus. In the human body, there are about 200 different types of cells (e.g., blood cells, liver cells, or lung cells), and a given virus usually is able to infect only a few of these many different cell types.

Viruses generally like to infect big organs. For example, the surface area of the respiratory tract is larger than a tennis court, so there are lots of cells in the respiratory tract for an inhaled virus to infect. The liver contains about one trillion cells, making this organ an attractive target. By infecting large organs which have many cells, a virus can kill or damage (either directly or indirectly) relatively large numbers of cells without doing serious harm to the human host.

Reproduction Within Host Cells

The second problem every virus must solve is how to reproduce within its target cells. No human virus carries with it the machinery (e.g., the ribosomes) required to synthesize proteins, and no human virus can generate the energy needed to power the copying (replication) of its genetic information. Because they lack the right stuff for their own reproduction, viruses must "hijack" some of the biosynthetic machinery of the cells they infect, and turn those cells into factories that can make many new copies of the virus.

Not only must a virus' genetic information be replicated, but viral proteins must be produced. Some viral proteins direct the replication process, and others are used in assembling the coat(s) that protect the viral genetic information once it leaves the cell. In actuality, a virus' reproductive strategy must have two parts: one for copying its genetic information, and another for producing the messenger RNA (mRNA) that will be translated to make viral proteins. Successful viruses have solved the reproduction problem so effectively that most can take over a human cell and use its biosynthetic machinery to produce thousands of new viruses.

Evasion of Host Defenses

Humans have evolved sophisticated mechanisms for detecting viral invaders and for dealing harshly with them. So the third problem every successful virus must solve is how to evade the host's antiviral defenses long enough either to infect another host, or to establish a **latent** or **chronic infection** within the original host—from which the virus can spread at a later time.

It is certain that as humans evolved new defenses, viruses were forced to evolve new evasion strategies. In fact, viruses and their human hosts are still involved in an ongoing struggle to gain the upper hand. It is important to understand, however, that viral evasion of host defenses need not be complete. In fact, there is a delicate balance here. If a virus is unable to evade host defenses sufficiently well, it will be destroyed before it can spread. On the other hand, if the virus is impervious to host defenses, it will likely kill the host before it can infect another person.

Transmission to New Hosts

The fourth problem every virus must solve is how to be transmitted from one infected individual to another. After all, even if a piece of genetic information evolves so that it can be replicated many times within a human cell, the wannabe virus will die when that human perishes—unless it can find a way to spread the infection to other humans. As you can imagine, this is not a simple problem for a virus to solve. First, newly made viruses must be transported out of the infected cell. Fortunately, cells have well established mechanisms that transport proteins made in the cell out to the cell surface. Consequently, by "borrowing" elements of the transport machinery of the cells they infect, newly minted **virus particles** usually can hitch a ride out of infected cells.

Once a virus has exited its infected cell, it must "arrange" to be physically transported from one host to the next. By taking advantage of human behavior as varied as coughing or having sex, viruses have evolved ingenious ways of establishing an unbroken chain of infection, thereby avoiding extinction.

CONSISTENT SOLUTIONS

So to be successful, every virus must solve four problems: how to infect its host, how to reproduce within its target cells, how to evade host defenses, and how to be transmitted to a new victim. Of course, the solutions to these problems must be consistent. For example, it wouldn't make much sense for a virus which can only reproduce in liver cells to be transmitted by coughing. Likewise, it wouldn't do for a virus to evolve to reproduce efficiently in cells of the intestine, yet not evolve a strategy that protects the virus from the acidic conditions present in the stomach, the gateway to the intestine. So it isn't good enough for a virus just to solve these four problems. The solutions must fit together in an overall plan of infection.

VIRAL PATHOGENESIS

The diseases which viruses cause are the result of the way each virus solves its problems of infection, reproduction, evasion, and transmission. Therefore, if we understand how a virus solves these four problems, we usually can predict what the pathological consequences of the viral infection will be. In some cases, the damage to the host that results from a viral infection is due to the actions of the virus itself (e.g., killing the cells it infects). In other situations, the host's reaction to the virus (e.g., the host immune response) can play a major role in the pathology. The reason for this is that most host defenses against viruses are not finely focused, so there is generally a lot of collateral damage. As my friend, Jim Cook, points out, the host's "shotgun" approach to defending against a viral infection is somewhat like trying to kill a mosquito with a machete: You may kill that mosquito, but most of the blood on the floor will be yours.

When you think about it, it really isn't in the best interest of a virus to seriously harm its host—in effect, biting the hand that feeds it. Indeed, many, perhaps most, of the pathological consequences of a viral infection are unintended.

FATHOMING THE MIND OF A VIRUS

I maintain that the goal in studying any virus should be to understand as fully as possible how the virus "thinks"—to view the world from the virus' perspective, not from the host's perspective (or the virologist's perspective!). The reason is that what a virus <u>does</u> will only make sense when we understand <u>why</u> the virus "chooses" to do things in that particular fashion. Because there are four problems which every virus must solve, our paradigm for fathoming the mind of a virus is to "ask the virus" the following questions:

1. How do you infect your host—and why do you choose this route?
2. How do you reproduce once you enter your target cells—and why do you choose this reproductive strategy?
3. How do you evade host defenses—and why do you employ these evasion tactics?
4. How are you spread to your next host—and why do you select this method of transmission?

The important point here is that once you have the virus' answers to these four questions, you will not just know <u>facts</u> about the virus—you will <u>know</u> the virus. I will illustrate the use of this paradigm by analyzing the lifestyles of 12 different viruses. I have chosen these viruses not only because they are important human pathogens, but also because their diverse ways of "doing business" serve to illustrate many general virological principles.

GENERAL PRINCIPLES

Throughout this text, we will encounter principles which apply to most, and in many cases all, viruses. I will list these at the end of the chapters in which we discuss them.

1. Viruses are pieces of RNA or DNA enclosed in a protective coat(s). These simple organisms are parasites which have evolved to reproduce inside, and survive outside, the cells they infect.
2. The viruses which now plague humans arose from within the cells that make up plants, humans, birds, and animals.
3. Every virus must solve four basic problems: how to infect its human host, how to reproduce within that host, how to evade host defenses, and how to spread the infection to the next host.
4. There are four major pathways to infection which viruses have evolved to use: Viruses can be inhaled, viruses can be ingested, viruses can be passed from mother to child, and viruses can be acquired by intimate physical contact.
5. Most viruses are very picky about which cells they infect — not just any cell will do.
6. A virus' reproductive strategy must have two parts: one for copying its genetic information, and another for producing the messenger RNA that will be translated to make viral proteins.
7. Viral evasion of host defenses need not be complete. The virus must only evade host defenses long enough either to spread to another host, or to establish a latent or chronic infection within the original host — from which the virus can spread at a later time.

Chapter 2

Host Defenses

BACKGROUND

The single most important factor which determines a virus' lifestyle is the constellation of host antiviral defenses it must protect itself against. We will examine some of the elegant strategies which viruses have evolved to evade host defenses. However, to fully understand how clever these viruses really are, we must first appreciate the power and diversity of the defenses arrayed against them. For this reason, before we begin the "interrogation" of our 12 model viruses, we will devote two chapters to an examination of host defense mechanisms.

Humans have three main types of defenses: physical barriers, the **innate immune system**, and the adaptive immune system. These defenses are arranged in "layers" so that invaders which penetrate one layer then must face the defenses located in the layer below. It is this multilayered defense which makes it possible for a human to survive the onslaught of tens of thousands of microbial invaders each day.

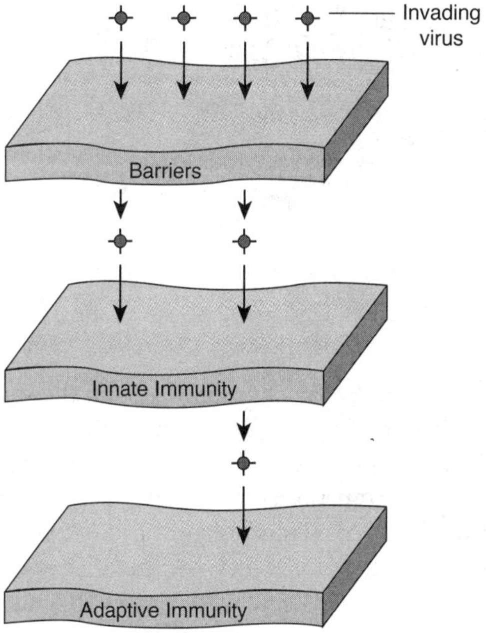

BARRIER DEFENSES

The first defense a virus must overcome is a physical barrier: the sheets of cells which cover the surface of our body, and which line its internal cavities.

The Skin

Of all the barriers, the skin is the most difficult for a virus to penetrate. The reason skin is such an effective barrier is that viruses can only infect living cells—and skin is covered with multiple layers of dead cells filled with keratin proteins. In fact, unless skin is punctured or torn, it is essentially impossible for a virus to use this route of infection. As a result, very few viruses have evolved to enter the body via the skin.

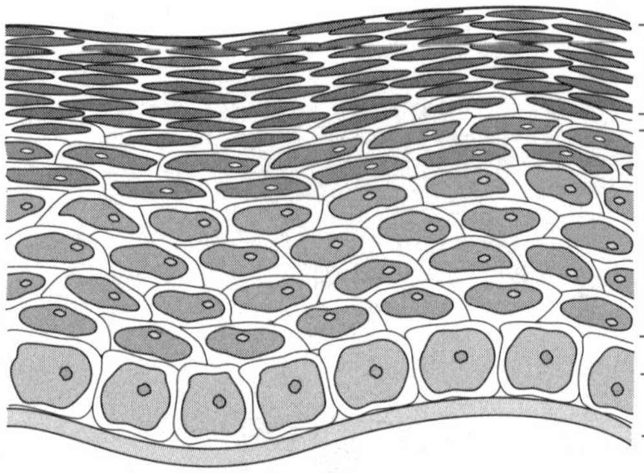

Most viruses have chosen to attack the mucosal barrier which lines our respiratory and digestive tracts. These surfaces must be permeable so that oxygen or nutrients can be transported across them. Consequently, parts of these mucosal surfaces are protected by only one or a few layers of living cells. Although a somewhat easier target than skin, mucosal surfaces still represent a formidable barrier to viral infection.

The Respiratory Tract

Much of the respiratory tract is lined with **epithelial** cells that have "paddles" (cilia) on their surface. These cells are also coated with a blanket of mucus. Indeed, each day about 100 mL of mucus is produced in the airway of a healthy human.

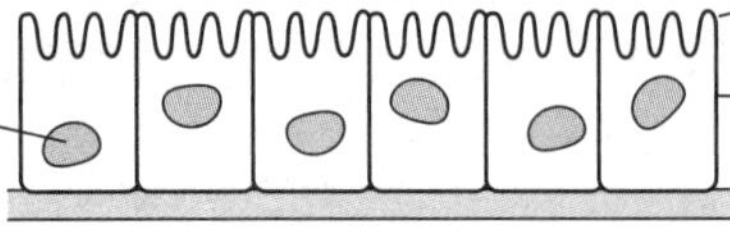

The cilia on these cells have a power stroke that is in the correct direction to move the mucus (and viruses trapped in it) toward the throat—where they can be coughed up or swallowed. Because they must swim against this current of sticky mucus, many viruses are unable to get a "grip" on their target cells before they are swept away. Actually "swim" is really the wrong word to use here, because no virus is able to swim. In fact, viruses are totally at the mercy of random motion to bring them close to their target cells so that they can bind to the appropriate receptor molecules on the cell surface. In addition to the mucosal barrier, some surfaces in the airway are protected by patrolling macrophages, white blood cells which can "eat" (phagocytose) viruses and limit their spread.

The Digestive Tract

Only the toughest of viruses would even consider invading via the digestive tract. To reach the intestine, where the cells they infect are located, these "**enteric**" viruses must first survive exposure to saliva which contains compounds that are active against many viruses. Indeed, although the mouth is quite accessible and seems like it should be an easy target, very few viruses cause infections of the oral cavity because of the potency of antiviral compounds in the saliva. For example, several proteins in saliva (e.g., secretory leukocyte **protease** inhibitor) are able to inactivate the AIDS virus, and this may explain, at least in part, why HIV-1 infections are not transmitted efficiently during oral sex.

After they battle their way past the saliva, enteric viruses next must brave the acid conditions of the stomach, where a pH as low as 2.5 is common. Also, they must avoid destruction by enzymes like pepsin that thrive at this pH, and which are tasked with helping break down proteins in food. Then, as they pass from the stomach into the beginning of the small intestine (the duodenum), viruses must contend with digestive enzymes that are piped into this region from the pancreas. These enzymes, which are designed to break down proteins, carbohydrates, and fats, really can do a number on the protein or lipid coat of most viruses. It is also here that bile salts from the liver are added to the digestive mix, and these act as detergents that help break up dietary fats—and some viral **envelopes**.

The Reproductive Tract

Whereas the respiratory and digestive tracts are mainly covered by an epithelial barrier that is only one or a few cells deep, the vagina is protected by an epithelium topped off by multiple layers (a "stratified" epithelium) of relatively flat (squamous), non-proliferating cells.

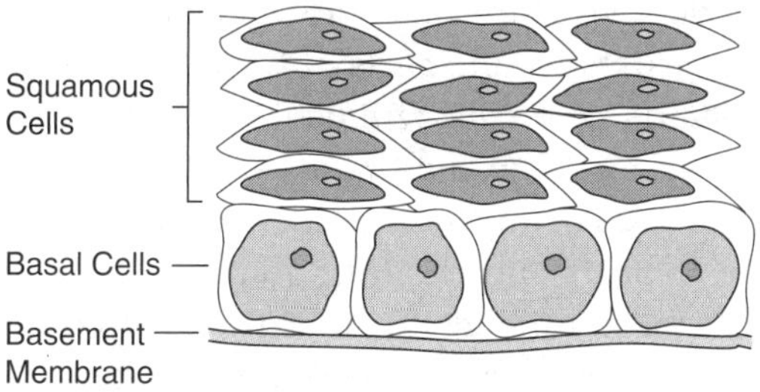

Viruses replicate best in human cells that are proliferating, so the multiple layers of non-proliferating cells in the stratified squamous epithelium of the vagina present a rather uninviting target for virus infection. Indeed, viruses which gain entry via the reproductive tract usually rely on small tears in the lining of the vagina (e.g., during sexual intercourse) to allow them access to the proliferating cells in the lower strata of the vaginal epithelium.

The epithelial cells which line the reproductive tract also are covered with mucus—which helps keep viruses at "arm's length." Moreover, the bacteria which colonize the vagina produce lactic acid, and this normally maintains the pH of the vaginal mucus around 5.0. Many viruses are sensitive to acid pH, so this acidic environment is yet another deterrent to viruses which might try to enter via the reproductive tract.

THE INNATE DEFENSE SYSTEM

Humans are vastly outnumbered by viruses, so the epithelial barriers cannot be expected to exclude every virus. Indeed, one of the important functions of the physical barriers is to decrease the number of invaders which must be handled by the next layer of defense—the innate immune system. This system is called "innate" because it is a defense that all animals seem to have.

The innate immune system includes three main weapons which can be brought to bear, singly or in concert, to defeat invading viruses: the professional phagocytes, the complement system of proteins, and the interferon warning system. Two of the qualities that make the innate system so important are that it responds quickly to a viral infection, and that it is "broad-spectrum"—being able to use the same weapons to defend against many different viruses. Here, we will discuss two of the three weapons of the innate immune system: the professional phagocytes and the complement system. The interferon system is so important for the host's defense against viruses that we will devote the entire next chapter to that system.

The Professional Phagocytes

Stationed beneath the sheets of epithelial cells are sentinel white blood cells. The most important of these "professional phagocytes" is the macrophage. These large cells are aptly named because they really are "big eaters" (phage is from a Greek word meaning "to eat"). Most of the time they patrol the tissues and "collect garbage," including debris from dead or dying cells. However, if they come upon a virus which has made it past the barrier defenses, they will eat it too. Sometimes this is a good thing, and the virus is destroyed by the battery of enzymes inside the macrophage. In other cases the virus uses the propensity of macrophages to ingest whatever they bump into as an easy way of infecting these professional phagocytes.

Most of the time, macrophages can deal with the occasional virus that slips through the barrier defenses, but sometimes the resident macrophages need reinforcements. Fortunately, the blood is teaming with professional phagocytes which can be summoned by embattled macrophages. The most numerous of these are the neutrophils—phagocytic cells which exit the blood, ready to kill. About 70% of all the circulating white blood cells are neutrophils—so there is plenty of backup available should the sentinel macrophages get in over their heads.

The Complement System

The complement system is composed of about 20 different proteins that work together to help destroy invaders and to signal other immune system warriors that the body is under attack. The complement system is very old. Even sea urchins, which evolved about 700 million years ago, have this system. The complement proteins are produced mainly by the liver, and are present at high concentrations in blood and tissues. The main function of the complement proteins is to bind to the surface of invaders, and tag them for destruction by professional phagocytes. Indeed, when a virus has its surface "decorated" with complement proteins, that virus becomes a much more attractive target for ingestion by macrophages and their relatives. In addition, complement proteins can bind to the surface of some viruses (e.g., HIV-1), and destroy these viruses by punching holes in their lipid coat.

THE ADAPTIVE IMMUNE SYSTEM

In some cases, the innate system is able to deal with viral attacks and destroy the invaders. However, the number of viruses present in the initial inoculum may be so large that the viruses, replicating to great numbers in infected cells, simply overrun the innate system. It is for this reason that humans and other higher vertebrates evolved another layer of defense: the **adaptive immune system**. The weapons of the adaptive immune system include antibodies (produced by B cells) and killer T cells.

B Cells and Antibodies

B cells are white blood cells that are born in the bone marrow, where they are descended from self-renewing stem cells. About one billion B cells are produced each day during the entire life of a human. B cells are "antibody factories" which make antibodies on demand. I say "on demand" because until they are alerted that there has been an invasion, B cells do not produce antibodies. Not only that, although there are enough B cells in a human to produce about 100 million different antibodies (each B cell makes only one kind of antibody), there are relatively few B cells that could produce any given kind of antibody. Consequently, B cells are essentially on "standby" until there is a viral infection. In the event of an attack, only those B cells which can recognize the virus are "activated." When this happens, the activated B cells begin to proliferate to build up their numbers, and then, finally, they begin to produce antiviral antibodies.

What Antibodies Do

Antibodies come in five "flavors" (classes): IgM, IgG, IgA, IgD, and IgE. Each of these classes has special properties, but only the first three classes are important in a viral infection. IgM and IgG antibodies are able to promote the attachment of complement proteins to the surface of viruses. So these antibodies function to direct the complement system to tag viruses so that the professional phagocytes can destroy them. IgM and IgG antibodies also can bind to viruses and "neutralize" them—either by preventing them from entering their target cells, or by disrupting viral reproduction once the virus has gained entry.

IgG antibodies are unique in that they can pass from the mother's blood into the blood of the fetus by way of the placenta. This provides the fetus with a supply of IgG antibodies to tide it over until it begins to produce its own—several months after birth. This "passive" immunity can protect the newborn against a variety of viral infections to which the mother has become immune because of previous exposures. IgG antibodies also are very important in protecting against viruses which travel via the bloodstream. Indeed, most of the antibodies in the blood are IgG antibodies.

If viruses invade parts of the body protected by mucosal surfaces, B cells will usually produce IgA antibodies—because IgA antibodies are just the ticket for defending against mucosal invaders. IgA antibodies are resistant to acids and enzymes found in the digestive tract, and these antibodies can be transported from the tissues, across the intestinal wall, and out into the intestine. There they can collect viruses together into clumps that are large enough to be swept out of the body with the mucus. The IgA class of antibodies also is found in the milk of nursing mothers. These antibodies coat the baby's intestinal mucosa and provide protection against pathogens that the baby ingests.

So, these different classes of antibodies are "specialists" which are particularly useful for defending against viruses which enter the body by different routes.

Killer T Cells

Although professional phagocytes and the complement system are quite effective against viruses which penetrate the physical barriers, these innate system weapons have one major flaw: The complement system and the professional phagocytes only have access to viruses which are outside of human cells. This is a big problem, because once a virus enters a cell, it is safe from these defenders—and an infected cell can produce thousands of new viruses in a period of a few days. What is needed is a weapon which can "look inside" cells, and efficiently destroy those which are being infected by a virus. That weapon is the killer T cell (frequently called a cytotoxic lymphocyte or CTL).

Killer T cells are white blood cells that are made in the bone marrow, but which mature in the thymus. Killer T cells have receptors on their surface which can recognize (bind to) proteins that are being displayed by class I major histocompatibility complex (MHC) molecules. Class I MHC molecules are "billboards" that exhibit on the surface of a cell, a "sampling" of all the proteins that are being manufactured inside that cell. This display includes ordinary cellular proteins like enzymes and structural proteins, as well as proteins encoded by viruses that are infecting the cell. Killer T cells use their receptors to "scan" the proteins displayed by the class I MHC billboards to determine whether a cell is infected. Most cells in the human body express class I MHC molecules, and are therefore an "open book" which can be checked by killer T cells. And when a killer T cell detects a virus-infected cell, it destroys it—and the viruses that are inside it.

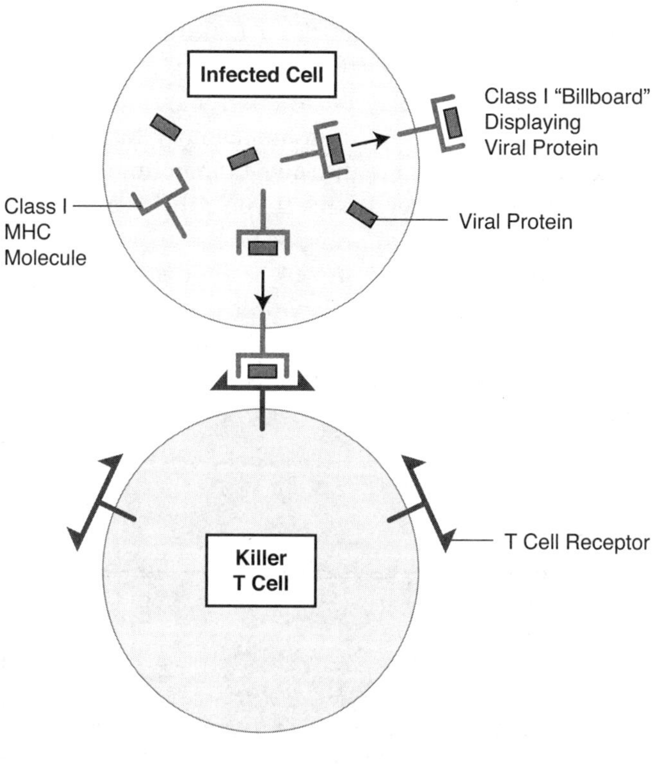

Each killer T cell has only one kind of receptor on its surface (i.e., a receptor which can recognize one particular viral protein displayed by an infected cell's class I MHC molecules). However, there are so many killer T cells in a human that the collection of T cells will include at least some with receptors that can recognize the proteins of any virus. Nevertheless, like B cells, there are very few killer T cells of any one kind. Consequently, T cells also are made on demand. As a result, when a killer T cell receives word that there has been an invasion, that T cell must first proliferate to make a large enough "clone" of killer T cells to mount a credible defense. And that takes a week or more.

From the virus' perspective, probably the most important feature of the antibody and killer T cell defenses is that during an initial attack, it takes a week or more for these weapons to be mobilized. This delay gives viruses "breathing room" to begin their infection without having to worry about being destroyed by the adaptive immune system—a fact that is critical for the survival of many viruses.

The Dendritic Cell

Residing in the tissues that underlie all the exposed surfaces of the body is a very important type of cell—the dendritic cell—sometimes called a "conventional" dendritic cell to distinguish it from other types of dendritic cells. It is this cell which is tasked with alerting the adaptive immune system when there is an attack. These sentinel cells are equipped with "sensor molecules" that can detect common signatures of most viruses. Interestingly, dendritic cells are "built" to be infected by viruses. They have a large collection of receptor molecules on their surface to which viruses can bind to gain entry, and their biosynthetic machinery is designed to allow the production of at least some viral proteins.

In response to the danger signals connected with a virus attack, dendritic cells do something very important: They exit the tissues where the battle is raging, and migrate through the lymphatic system to nearby lymph nodes. There, the dendritic cell sounds the alert so that B cells and T cells can be activated.

Helper T Cells

Dendritic cells that travel from the battle scene to lymph nodes do something else which is important: They activate helper T cells. In contrast to killer T cells, helper T cells don't kill—they "assist" other immune system cells. For example, helper T cells provide confirmatory signals which B cells need to begin antiviral antibody production. Helper T cells also give off chemical messenger molecules (cytokines) which help determine the class of antibody B cells will produce, and which help keep killer T cells activated and ready to go.

Dendritic cells and helper T cells are essential components of the adaptive immune defense. Without them, the system doesn't function. That is one reason why an infection with the AIDS virus can be so devastating: HIV-1 targets and destroys dendritic cells and helper T cells, leaving the infected human with a profoundly impaired adaptive immune system.

Memory B and T Cells

During a viral infection, some B cells "choose" to become "memory" B cells instead of becoming antibody factories. These memory cells are set aside to protect us against another attack by the same virus. Because these cells have already "seen" the virus, they have properties that make them much better able to respond to a subsequent attack. First, there are many more of them, so relatively little **proliferation** is required to produce enough of these cells to be effective. Second, they have been "trained" to produce the type of antibody (e.g., IgG or IgA) which is appropriate for the virus they remember. Finally, their detectors have been fine-tuned so that they can be activated very early in an infection when only tiny amounts of the attacking virus are present. The result is that memory B cells can be mobilized rapidly, and can produce huge quantities of antibodies which are exactly right to defend against that particular virus. This is the reason antibodies usually play the major role in protecting us against subsequent infections by the same virus. Indeed, it is the goal of most vaccination strategies to produce memory B cells and protective antibodies.

Memory killer T cells also can be produced during a viral infection. Like memory B cells, there are more of them than were present at the time of the original infection, and they are easier to mobilize. Consequently, memory killer T cells provide a faster response to a subsequent infection by the virus they remember.

Some viruses reproduce so rapidly that the long delay in mobilizing the adaptive immune system means that the initial infection is essentially a "freebie"—at least as far as the adaptive immune system is concerned. These viruses can reproduce and "leave the building" before antibodies and killer T cells can be deployed. However, the ability of the adaptive immune system to remember the first attack and be ready for a subsequent invasion has serious implications for many viruses. Indeed, the existence of immune memory can make a successful return visit to the same human impossible.

In this chapter I have given a brief overview of host defenses which are specific for viruses. If you would like a more in-depth look at these defenses, or if you would like to see how they protect us against other attackers (e.g., bacteria), you may wish to read one of my other books, *How the Immune System Works* (4th ed., Wiley-Blackwell, 2012).

GENERAL PRINCIPLES

1. The physical barriers which separate our bodies from the outside world greatly decrease the number of invaders which must be handled by the next layer of defense – the innate immune system.
2. The innate immune system includes three main weapons which can be brought to bear, singly or in concert, to defeat invading viruses: the professional phagocytes, the complement system of proteins, and the interferon warning system.
3. Two of the qualities that make the innate system so important are that it responds quickly to a viral infection, and that it is "broad-spectrum" – being able to use the same weapons to defend against many different viruses.
4. The complement system is composed of about 20 different proteins that work together to help destroy invaders and to signal other immune system warriors that the body is under attack.
5. The different classes of antibodies are "specialists" which are particularly useful for defending against viruses which enter the body by different routes.
6. When a killer T cell detects a virus-infected cell, it destroys the cell – and the viruses which are inside it.
7. Many viruses reproduce so rapidly that the long delay in mobilizing the adaptive immune system means that the initial infection is essentially a "freebie," at least as far as the adaptive immune system is concerned. These viruses can reproduce and "leave the building" before antibodies and killer T cells can be deployed.
8. Memory B and T cells are "upgraded" versions of the weapons that responded to the original infection. Consequently, they are better able to defend us against a subsequent attack by the same virus.

Chapter 3

The Interferon Defense System

BACKGROUND

Although the complement system and the professional phagocytes are important in limiting the intensity of a viral infection, the most critical early host defense against a viral attack is the interferon system. Indeed, the interferon system is the defense which viruses fear most. Every virus has evolved multiple ways to fend off this system—at least long enough to reproduce and spread to a new host. Consequently, if we wish to understand how viruses "think," we must understand the interferon system and the deadly threat it poses to viruses.

VIRAL DETECTION

Human cells have evolved sensor molecules which can detect molecular signatures (patterns) that are common to different classes of viruses (e.g., viruses that generate double-stranded RNA during replication). Once these receptors detect such a viral signature, they initiate a protein **kinase**–signaling cascade which conveys the "under attack" message to the cell nucleus, and turns on the production of "warning proteins" called type 1 interferons: interferon-α (IFN-α) and interferon-β (IFN-β). In a human cell, there is one gene for IFN-β, and many genes for slightly different forms of IFN-α.

Double-Stranded RNA

An excellent example of a cellular **pattern recognition receptor** is the "**Toll-like receptor**," TLR3, which can detect long stretches (generally longer than 30 base pairs) of double-stranded RNA inside virus-infected cells. Normal cells contain only small amounts of double-stranded RNA, and it usually is relatively short. In contrast, many viruses produce lengthy stretches of double-stranded RNA when they replicate. When TLR3 detects this "molecular pattern," it initiates a kinase cascade which phosphorylates a protein called interferon regulatory factor 3 (IRF3). In most cells, IRF3 is already present in an inactive, unphosphorylated form—so it is ready to go once it is phosphorylated. When this happens, the phosphorylated IRF3 enters the nucleus of the cell, and acts as a transcription factor to turn on expression of the IFN-β gene.

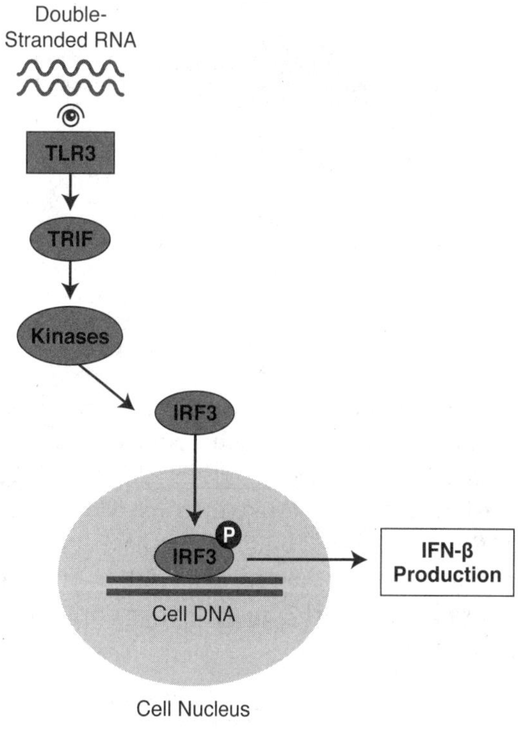

Abnormally Terminated RNA

Another pattern recognition receptor, RIG-I, detects single-stranded RNA which has a 5′ triphosphate—a type of RNA synthesized by the polymerases of many viruses. Normal cellular RNAs either are capped at their 5′ end (mRNAs) or have a 5′ monophosphate (ribosomal RNA and tRNA). Consequently, the presence of an RNA molecule with a 5′ triphosphate is a "danger" signal that can be detected by the RIG-I sensor. When this happens, RIG-I initiates a kinase cascade that is similar to the one initiated by TLR3. The result is that IRF3 is phosphorylated, allowing it to turn on expression of the gene which encodes IFN-β.

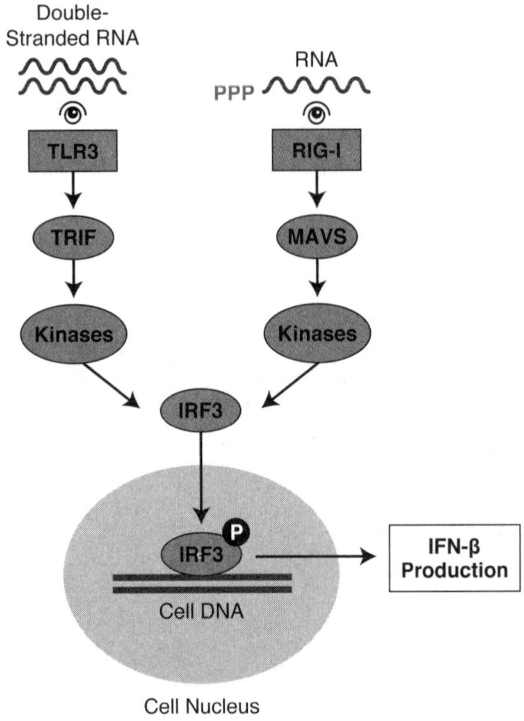

Retroviral RNA

Retroviruses (e.g., HIV-1) are single-stranded RNA viruses which replicate through a double-stranded DNA intermediate. When retroviruses invade cells, their RNA is sensed by another Toll-like receptor, TLR7. When retroviral RNA is detected by TLR7, a kinase cascade is initiated that results in the phosphorylation of interferon regulatory factor 7, IRF7, and the transcription of both IFN-α and IFN-β genes.

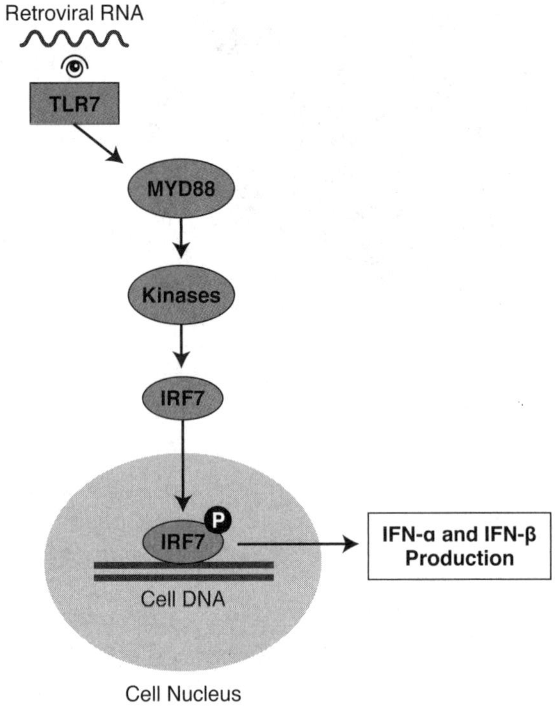

Unmethylated Dinucleotides

Certain viruses with double-stranded DNA genomes (e.g., herpes simplex virus) can be detected by TLR9, which senses stretches of unmethylated CpG dinucleotides that are rare in cellular DNA. When this "unusual" DNA is detected by TLR9, a kinase cascade similar to that triggered by TLR7 is initiated, resulting in the turn-on of the genes for IFN-α and IFN-β.

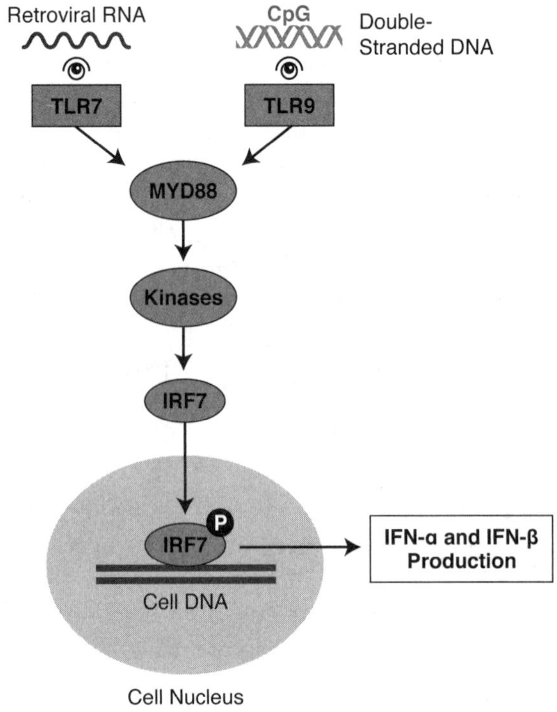

Other detectors of common viral signatures have been identified, and more will likely be discovered in the future. The important point here is that these pattern recognition receptors can discriminate between normal cellular molecules and molecules that are uniquely viral, and can trigger the production of interferon.

INTERFERON FUNCTION

Once IFN-α or IFN-β has been produced, it is transported out of the infected cell and can bind to interferon receptors on the surface of nearby cells. Type 1 interferons also can bind to receptors on the surface of the cell which produced them. The function of the interferon proteins is to orchestrate a series of events which "interfere" with viral reproduction. One way interferons do this is by alerting nearby uninfected cells that viruses are in the area, and that they may soon be attacked. IFN-α and IFN-β bind to the same receptor, and most cells in the human body have interferon receptors—so this warning system is quite general.

Warning Nearby Cells

When interferon engages a cell's interferon receptors, a signal transduction pathway is activated that results in the formation of a complex of proteins (including STAT1, STAT2, and IRF9) called interferon stimulated gene factor 3 (ISGF3). This complex of proteins enters the cell nucleus and functions as a transcription factor to turn on the expression of literally hundreds of genes—the interferon stimulated genes (ISGs). The proteins encoded by many of these ISGs have antiviral activities.

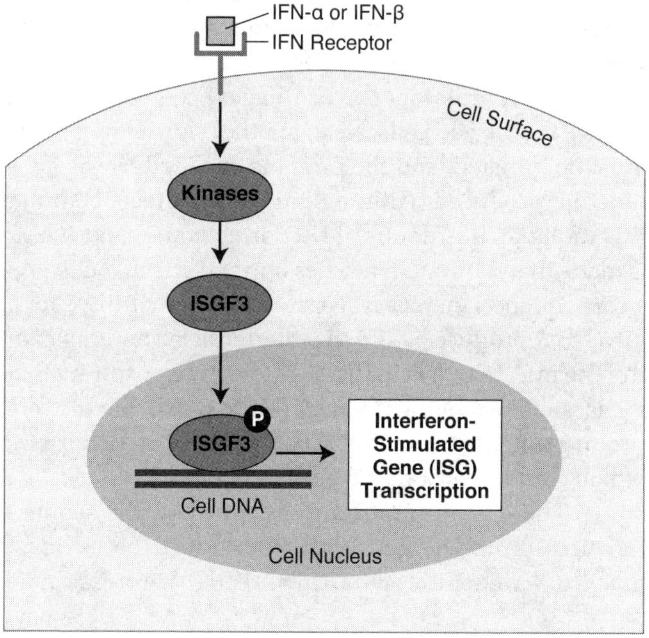

One well characterized interferon stimulated gene encodes a protein kinase called PKR. When interferon binds to its receptors, the cell begins to produce large amounts of this protein. The PKR protein has the remarkable property that it can sense the presence of double-stranded RNA, and can shut down protein synthesis if the "warned" cell is subsequently infected with a virus which produces double-stranded RNA. Here's how this works.

When PKR binds to double-stranded RNA, it is activated, and in its activated state, it can attach phosphate molecules to eukaryotic initiation factor 2 (eIF2), a protein required to initiate cellular protein synthesis. However, when eIF2 has those extra phosphates attached, it can no longer function. Consequently, when a virus infects an interferon-warned cell and produces double-stranded RNA, PKR is activated, eIF2 is phosphorylated, and protein synthesis ceases. Because both the infected cell and the virus need protein synthesis to survive, this altruistic defense leads to the death of the infected cell—and the virus trying to reproduce within it.

Another interferon warning system involves two other ISGs, the enzymes 2′-5′-oligo(A) synthetase and RNAse L. Like PKR, the expression of theses two proteins increases dramatically when interferon binds to a cell's receptors. Also like PKR, the 2′-5′-oligo(A) synthetase protein is activated by double-stranded, viral RNA. And once this enzyme becomes active, it can activate RNAse L, which in its active form can degrade both host and viral RNA, halting protein synthesis.

The elegant part of the interferon warning system is that although the binding of interferon to its receptors on an uninfected cell prepares the cell for a viral attack by upregulating the expression of interferon stimulated genes, that cell continues to do business as usual unless an attack actually occurs. Only if a virus tries to infect the cell will the defense be activated. Moreover, if the attack does not come, the warned cell eventually "stands down" from its state of readiness. Although the "antiviral state" is temporary, the interferon defense is quite effective in preventing viruses from infecting other cells.

Interfering With Reproduction in Infected Cells

Although one effect of interferon production by a virus-infected cell is to warn nearby cells of an impending attack, there is another function of interferon. Because interferon receptors also are found on the surface of the original, infected cell, the interferon produced can "feed back" on that cell, inducing the expression of its own interferon stimulated genes. This can be especially bothersome for viruses which reproduce slowly, because the interferon "warning" system can limit viral synthesis in the original infected cell.

In summary, when pattern recognition receptors sense the presence of a virus, they trigger the production of interferon, which activates expression of interferon stimulated genes (ISGs)—resulting in the synthesis of proteins that can function to interfere with viral reproduction. Amazingly, although ISGs are extremely important in protecting us against viral infections, very little is known about how most of them function. Clearly there is still a lot which virologists don't understand!

SEQUENTIAL PRODUCTION OF IFN-β AND IFN-α

In most cells in the body, viral detection by pattern recognition receptors initially turns on the expression of IFN-β, but not IFN-α. The reason for this is that IRF3 is already present in most cells in an inactive form when a virus attacks, and only needs to be phosphorylated to turn on IFN-β production. In contrast, the transcription factor, IRF7, which is required to activate the genes for IFN-α, normally is not produced in uninfected cells. However, once IFN-β is produced, it is exported, binds to interferon receptors, and initiates a signaling cascade which turns on interferon stimulated genes. Importantly, one of these ISGs is the gene for IRF7. And once IRF7 has been produced, it can be phosphorylated by the same signaling cascade that activates IRF3; be translocated to the cell's nucleus; and turn on not only the gene for IFN-β, but also the genes for IFN-α. Now the interferon system is at full strength to defend against the virus attack.

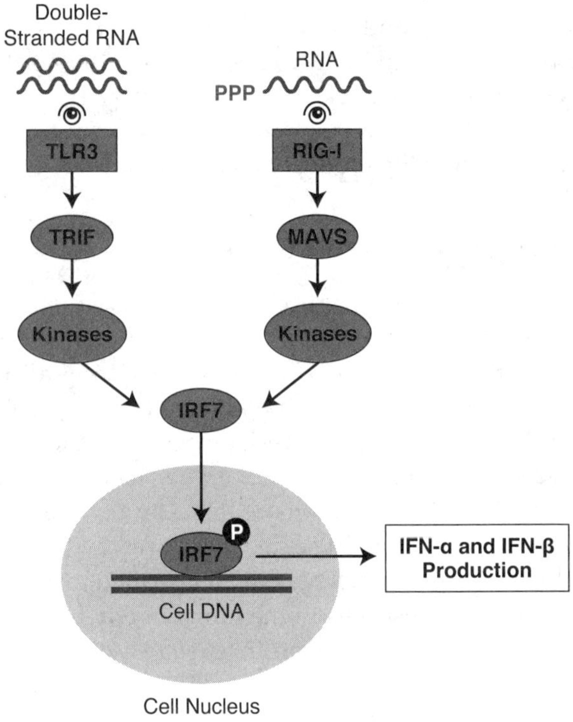

So in most cells, the interferon defense begins with the expression of IFN-β, which then, indirectly, turns on the expression of the IFN-α genes to amplify the interferon response. The fact that IFN-β and IFN-α are turned on sequentially has practical importance for an attacking virus: Although the interferon defense is activated quickly, it is not instantly available. There is a delay. When the virus is detected by pattern recognition receptors, IRF3 must be activated, and the IFN-β protein must be made. This first step in turning on the system usually takes about six hours. Once IFN-β has been produced, it is transported out of the cell, and binds to interferon receptors on the infected cell itself and on nearby, uninfected cells. Next, another signaling cascade must be initiated, and the proteins encoded by the ISGs must be produced. Although antiviral effects begin at this moment, it is not until IRF7 is activated and IFN-α is made that the system reaches optimal effectiveness. As a consequence, there is a "window of opportunity" for a virus to reproduce before the interferon system can control the infection. It is in this window of opportunity that viruses make a living. Indeed, if IFN-β and IFN-α were produced instantly, there probably would be no viruses which could persist in the human population.

There is an important exception to the sequential turn-on of the genes for IFN-β and IFN-α. Plasmacytoid dendritic cells (pDCs) are "professional" interferon-producing cells which can synthesize up to 1,000 times more interferon than any other cell in the body. One of the things which makes these cells exceptional is that uninfected pDCs produce large amounts of the IRF7 transcription factor, so this protein is ready and waiting to be activated. Moreover, these cells are equipped with TLR7 and TLR9—Toll-like receptors which can activate IRF7 in response to infection by retroviruses such as HIV-1 (via TLR7) and certain double-stranded DNA viruses (via TLR9). Consequently, in response to an attack by these viruses, plasmacytoid dendritic cells bypass the requirement to produce IFN-β first, and begin to make both type 1 interferons at the same time.

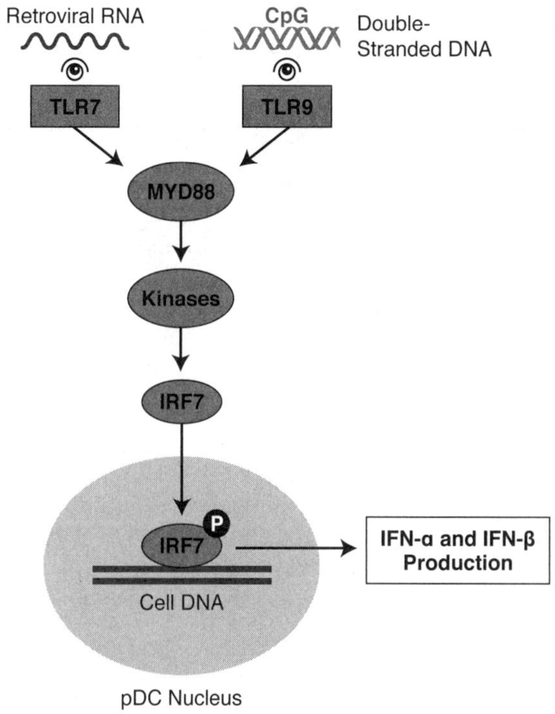

VIRAL EVASION OF THE INTERFERON DEFENSE

Every virus which has been studied activates the interferon defense system, and every successful virus has evolved multiple ways to subvert this defense. Indeed, as more and more is discovered about the interferon system, it becomes increasingly clear that, over evolutionary time, viruses and cells have been locked in a life-and-death struggle to refine and manipulate this system. As we examine the 12 viruses in our Parade of Viruses, we will see many examples of the ways viruses have "learned" to evade the interferon defense.

GENERAL PRINCIPLES

1. The most critical early host defense against a viral attack is the interferon system.
2. Pattern recognition receptors can discriminate between normal cellular molecules and molecules that are uniquely viral, and can trigger the production of interferon.
3. Once IFN-α or IFN-β has been produced, it is transported out of the infected cell and can bind to interferon receptors on the surface of nearby cells. These type 1 interferons also can bind to receptors on the surface of the cell which produced them.
4. When pattern recognition receptors sense the presence of a virus, the end result is the production of interferon and the activation of interferon stimulated genes (ISGs) — which encode proteins that can interfere with viral reproduction.
5. Although the interferon defense is activated quickly, it is not instantly available. There is a delay. Indeed, if IFN-β and IFN-α were produced immediately, there probably would be no viruses which could persist in the human population.
6. Every virus which has been studied activates the interferon defense system, and every successful virus has evolved multiple ways to subvert this defense.

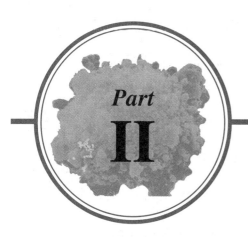

Part

II

The Bug Parade

Now we come to what my friend, Tom Hill, calls the "Bug Parade" — the part where we meet our 12 model viruses. Most of the time, viruses are grouped according to their family classification. In this book, however, I'm going to parade the viruses according to the route they use to enter the body — because the route of infection is so important in determining a virus' lifestyle. Although viruses sometimes are versatile enough to use several ports of entry, I will position each virus in the Parade according to what I consider its most "natural" route of infection. For example, a particular virus might be spread either during the birth of a child (a very natural route, you must agree) or by the sharing of contaminated needles. Such a virus would be grouped with "Viruses We Get From Mom," since it is clear that none of these viruses evolved to use drug abuse as their natural mode of transmission.

As the 12 viruses march by, each one will be taken aside for a brief "interview." You will notice that the "Interviewer" uses the paradigm introduced in Chapter 1 to try to understand life from the virus' point of view. In each case, he will ask the virus how it deals with the four problems every virus must solve:

1. How do you infect your host — and why do you choose this route?
2. How do you reproduce once you enter your target cells — and why do you choose this reproductive strategy?
3. How do you evade host defenses — and why do you employ these evasion tactics?
4. How are you spread to your next host — and why do you select this method of transmission?

As you read these interviews, I'm sure you'll join me in marveling at the elegant strategies that viruses have evolved to solve their four problems. Let the Parade begin!

Chapter 4

VIRUSES WE INHALE

Influenza: A "Bait-and-Switch" Virus

BACKGROUND

There are three types of human influenza viruses: A, B, and C. Influenza A is the most dangerous of these, killing roughly 500,000 humans each year, so we certainly need to try to fathom the mind of this virus. Of course, influenza B and C viruses also have features which make them interesting, especially when their properties contrast with those of influenza A. All three types can cause typical flu symptoms, yet these viruses are sufficiently different that antibodies against one type will not protect against the others. Strains of type A influenza infect animals as diverse as birds, pigs, seals, horses, and ferrets. Type B and type C influenza viruses are primarily human viruses, although they occasionally have been isolated from animals, but not from birds. In this interview, we will focus on type A influenza.

INFLUENZA VIRUS INFECTION

Interviewer: One question I always ask the viruses I interview is: How do you attack your hosts, and why have you chosen that route? I think this is an important question, because how a virus enters a human and what cells it infects is a critical determinant of its lifestyle.

Flu virus: I favor the respiratory route.

I: Okay, but why? For example, why not enter via the digestive tract?

FV: Are you kidding me? Do I look like a dumb virus to you? My Uncle Harold tried the digestive tract once, and got as far as the stomach before the acid in there ate him alive! Not me. I take the easy way in. The respiratory route of infection provides direct access to my favorite target cells—the epithelial cells which line the airway. In addition, I usually spread from human to human in microdroplets created by a sneeze or cough of an infected individual. Because of their small size, when these microdroplets are inhaled by my next host, they can penetrate all the way to his lungs. As a result, I can infect epithelial cells that line both the upper and the lower respiratory tract, making the entire airway—all 1,000 square feet of it—my playground! And another thing. Once I have infected an epithelial cell, it only takes me a few hours to produce thousands of new viruses—viruses which then go on to infect other cells of the respiratory tract. In fact, I can usually infect almost every cell in a given branch of the airway.

I: I'm impressed. But the epithelial cells which line the upper portion of the respiratory tract have "paddles" (cilia) on their surfaces, and these cells are coated with a blanket of mucus. Doesn't that make entry by the respiratory route difficult?

FV: Yeah, those cilia are a problem. They have a power stroke that is in the correct direction to move the mucus (and viruses like me!) toward the throat, where we can either be coughed up or swallowed. And I really hate that mucus! How would you like to try to "swim" against a current of sticky mucus, and be unable to get a grip on your target cell before you're swept away! That's no fun. And then there are those vicious macrophages. The lower part of the respiratory tract (the alveoli) is patrolled by macrophages—killing machines whose sole purpose in life seems to be to "eat" (phagocytose) viruses like me. It's a dangerous world down there!

I: The respiratory route doesn't sound all that easy to me. However, you are a very successful virus, so you must have some way of overcoming those obstacles and reaching your target. What's your secret?

FV: Numbers. Sheer numbers. A hearty sneeze can expel more than 10,000 influenza virus–containing droplets, making it likely that I, or some of my relatives, will penetrate the mucus, avoid the macrophages, and infect the underlying epithelial cells. They may stop some of us—or even most of us—but they can't stop us all!

I: Because viruses must use a cell's biosynthetic machinery, all viruses must somehow get their genetic information across the cell's plasma membrane and into their target cell. The first step in this process involves binding of the virus to receptor molecules on the cell surface. How do you accomplish this?

FV: It's not a problem. My RNA genome is coated with virus-encoded proteins, and enclosed in a patch of membrane (my "envelope") which I picked up from the last cell I infected. On the outside of my envelope are about 500 copies of a protein called the viral hemagglutinin. This protein binds to sialic acid (a.k.a. neuraminic acid) residues of proteins on the surface of my target cells.

You might be interested in a bit of history. The hemagglutinin protein got its name when it was discovered that when virologists mix influenza viruses with red blood cells (which contain heme), the blood cells clump (agglutinate). This happens because there are proteins with sialic acid residues on the surface of red blood cells. Consequently, the hemagglutinin proteins on the envelope of an influenza virus can form "bridges," connecting many red blood cells together. Different influenza virus strains can have slightly different hemagglutinin molecules. As a result, the type of cell or the species we infect (our **"host range"**) is limited by the ability of our viral hemagglutinin proteins to bind to the different types of sialic acid–containing molecules available on the surface of various host cells (e.g., human versus bird).

I: Interesting history, but many cells in the human body have sialic acid residues on their surface, and yet you seem to specialize in infecting cells in the respiratory tract. So there must be more to your host range than just those receptor molecules.

FV: Yes, indeed there is. It turns out that to succeed in entering a cell, my hemagglutinin protein must first be cut into two pieces. This cleavage can be carried out either by enzymes that are inside a cell in which new viruses are being produced, or later by enzymes that are in the environment outside the cell. Only certain kinds of cells produce enzymes which can do this cutting, and only certain cellular environments contain the needed enzymes. As a result, the requirement for hemagglutinin cleavage can limit the types of cells I can infect. Most human cells, including airway epithelial cells, lack the enzyme (a serine protease) required to cleave the hemagglutinin proteins of the strains of type A influenza which normally infect humans. This means that the all-important cleavage step must be performed after newly made viruses have been released into the respiratory tract. Fortunately for me, there are cells in the airway (e.g., Clara cells) which produce serine proteases and export them out into the mucus. Consequently, I take advantage of the enzymes present in the mucus that coats the cells of the respiratory tract to cleave my hemagglutinin molecules to a functional form. Come to think of it, I probably shouldn't have bad-mouthed that mucus. I really couldn't live without it. But it's a pain to get through.

I: Once a virus binds to its target cell, it is faced with the task of entering the cell and removing its protective coat—so that its genetic material can be replicated. What interests me about this process is that the viral coat must be stable enough to protect the viral genome from the environment that surrounds the cell, yet once the virus has entered, the coat must be at least partially disassembled to allow replication to proceed. How do you solve this "uncoating problem"?

FV: Viruses employ two general strategies for entry and uncoating. In the first, the virus binds to its receptor on the cell surface, "checks its overcoat at the door," and injects its partially uncoated genome into the cell's cytoplasm. However, my relatives and I use a different technique to enter a cell: **receptor-mediated endocytosis**. During this process, we first bind to sialic acid residues on the cell surface. Then, as a result of this binding, the cell encloses us in a portion of its plasma membrane (called an **endosome**), and takes us right inside. It's almost too easy: We just knock on the door, and the cell invites us in!

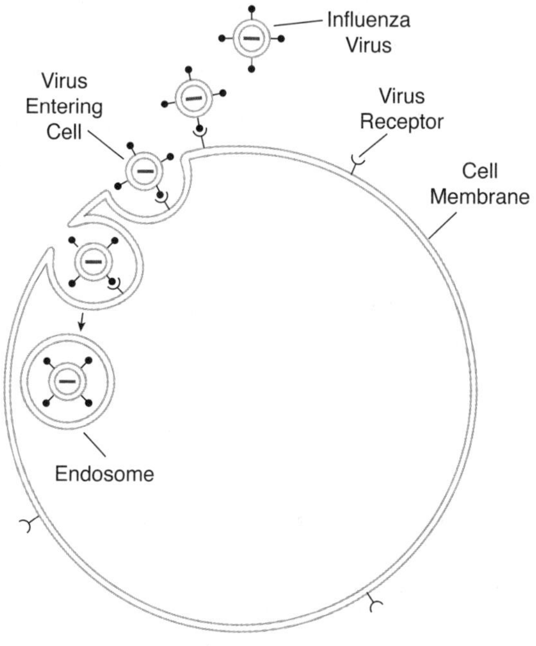

I: So receptor-mediated endocytosis gets you into the cell, but you still must figure out how to shed your protective envelope and get your genome out of the endosome.

FV: Got that covered. The plasma membrane of a cell contains "pumps" which normally transport protons out of the cell. However, when the endosome is formed, the membrane is inverted, so that protons are now pumped into the interior of the endosome. As a result, the environment within the endosome becomes progressively more acidic. When the pH inside of the endosome reaches about 5.0, a conformational change takes place in my envelope which allows it and the endosome to fuse, releasing my RNA genome into the cytoplasm.

INFLUENZA VIRUS REPRODUCTION

I: Pretty slick. So now you have made it into the cytoplasm where your goal is to make many more influenza viruses. What next?

FV: My genome is composed of eight short pieces of single-stranded RNA which encode a total of 11 proteins. Once released into the cytoplasm, these RNA segments rapidly enter the nucleus of the cell, being directed there by nuclear localization signals present on the viral proteins that remain bound to the segments of RNA. It is in the nucleus of the infected cell that replication of my genome takes place. I must point out that this is quite unusual for a virus with an RNA genome. Almost all the rest of them replicate in the cytoplasm. But then, influenza viruses dare to be different.

I: The genomes of single-stranded RNA viruses can be either "positive-strand" or "negative-strand." Positive-strand viral RNA is defined as RNA which is ready to be translated into proteins—so it is synonymous with viral messenger RNA. Negative-strand RNA is the "opposite" strand, so a complementary copy of it must be made to obtain viral messenger RNA. Which are you?

FV: I'm a negative-strand RNA virus—and proud of it. Did you know that human cells don't even have polymerase enzymes which can replicate RNA molecules? Consequently, every virus that has an RNA genome must encode its own polymerase. The RNA of negative-strand viruses like me cannot be translated into proteins, so we must carry our personal polymerase proteins along with us inside our protective coats. During an infection, one of my polymerase molecules remains associated with each of my eight RNA segments. That way, once I reach the cell's nucleus, my polymerases can spring into action, and can synthesize complementary copies of my viral RNAs (vRNA) to yield positive, protein-coding, messenger RNA (mRNA).

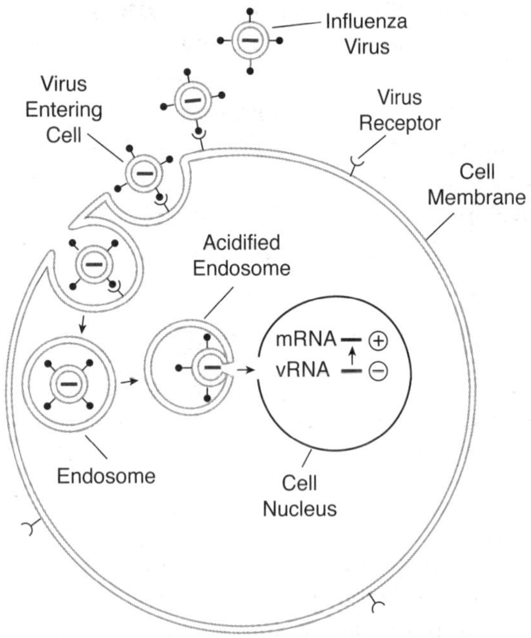

These messenger RNA molecules are then transported out into the cytoplasm, and are translated on cellular ribosomes to make new viral proteins, including more polymerase molecules. The newly synthesized viral polymerases then re-enter the nucleus where they produce more complementary copies (cRNAs) of the original viral RNAs. These positive-strand cRNAs are then recopied many times by these viral polymerases to make the negative strands that will form the genomes of newly minted viruses. It sounds complicated—all this in and out—but it's really pretty simple.

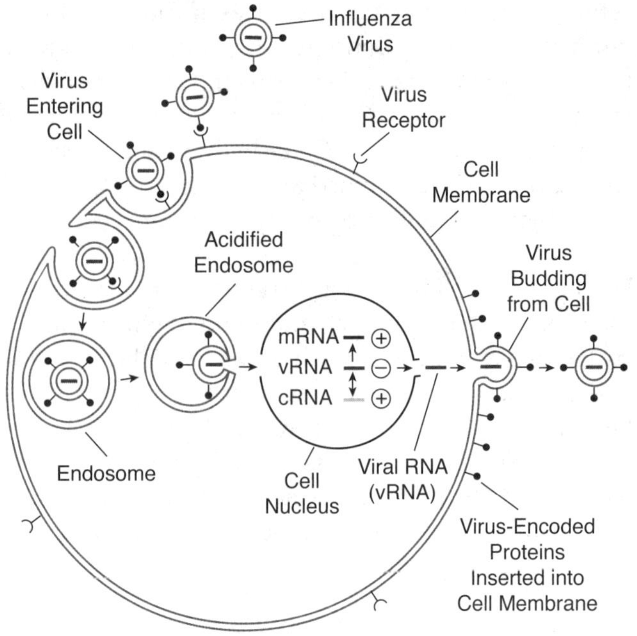

A Dirty Trick

I: I've heard that you do something in the cells you infect which sounds rather—how shall I put it—nasty. Something called "cap snatching." What's that about?

FV: Yes, I've gotten a lot of bad press about this one, but let me explain. To be translated by cellular ribosomes, my mRNA must have a "cap" structure at its 5′ end. I didn't make this rule. That's just the way it is. Now I could either go to the trouble of encoding all the various enzymatic activities required to synthesize a cap for my mRNAs, or I could just arrange to bite the cap off of a cellular mRNA molecule, and paste the cap onto my mRNA.

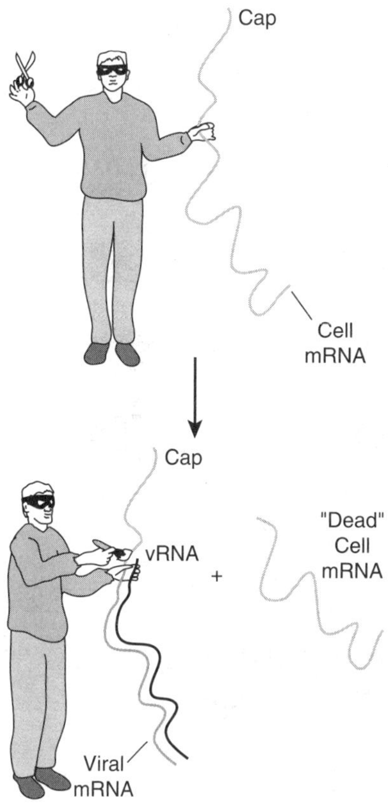

I'm just a tiny little virus. I only have the coding capacity for 11 proteins. So I said to myself, "Why not get a cap for free?" It's so easy. But there's another reason I snatch caps. My basic strategy is to reproduce quickly and get on to my next victim—sorry, my next "host." In the cells I infect, there are already hundreds of thousands of mRNA molecules—all competing for ribosomes. What chance do I have to get my viral proteins made quickly with all this cellular protein synthesis going on? No chance at all. So the second reason I snatch caps is that without their caps, those cellular mRNAs can't be translated. So by snatching the caps from cellular mRNAs, I focus the protein-making machinery of the cell on making the viral proteins I need to reproduce quickly. As a result, my viral replication cycle takes only about six hours to complete. You can ask any virologist and he'll tell you that's really fast! The sad part of all this, however, is that, as a consequence of my rapid reproduction style, infected cells die shortly after newly made viruses are released. Viruses like me, which kill the cells we infect, are called "**cytolytic**" viruses. Yes, it's a dirty trick. I'm not proud of it. But a virus has gotta do what a virus has gotta do.

Exit Strategy

I: I suppose so. But let's go on. After you have forced the cell to make many new viruses, you must somehow arrange to transport these viruses out of the cell. What's your "exit strategy"?

FV: After newly synthesized viral RNA segments (vRNAs) are coated with virus-encoded proteins, they exit the nucleus, and proceed to the inner surface of the plasma membrane. There they wait until all eight gene segments have been gathered together. Finally, the whole collection of protein-coated vRNAs "buds" from the cell surface, picking up a patch of cell membrane which forms my viral envelope. Don't ask me how I arrange to collect all eight of the gene segments I need for a complete genome. Even the smartest virologist doesn't know the answer to that question! It's one of my secrets. What I will tell you, however, is that, prior to budding, several viral proteins (including the viral hemagglutinin) are inserted into the cell membrane. Consequently, my envelope contains viral proteins in addition to the normal cell membrane components. I'm not unique in this respect: Enveloped viruses always arrange to have viral proteins inserted into the cell membranes through which they bud. They have to, because one or more of these viral proteins will form the "plug" that will engage the "socket" (receptor) on the surface of the cell the virus wants to infect next. No plug, no infection.

I: I understand that having hemagglutinin proteins attached to the outside of your viral envelope gives you a way of engaging your next target, but doesn't this also pose a problem? After all, the cell you infected has proteins on its surface that contain the sialic acid molecules to which the hemagglutinin protein binds. That's how you entered the cell in the first place. So why aren't newly minted viruses "recaptured" by the infected cell when they try to leave?

FV: Unfortunately, some of those young viruses are recaptured. However, to minimize this annoying problem, we influenza viruses produce a second protein which, like the hemagglutinin protein, is inserted into the cell membrane during infection. This "neuraminidase" protein functions as a "razor" that enzymatically shaves off sialic acid residues from the surface of the infected cell. Once these sialic acid molecules have been removed, we can bud through the "shaved" membranes and escape, unencumbered by those sticky receptors.

EVADING HOST DEFENSES

I: Very clever. But now these newly made viruses must face host defenses. How, for example, do you deal with the interferon system?

FV: Interferon really frightens me! And not just me. The innate defense which all viruses fear most, especially during their first visit to a human, is the interferon system. We viruses have been engaged in an "arms race" with the interferon defense for a very long time. And just when we get the upper hand, humans come up with a new twist on the interferon system which we must then try to deal with. It's very frustrating—and scary. Through all this, we viruses have evolved three main strategies for protecting ourselves from the horrors of the interferon defense. The first is to try to block the production of interferon. Our thinking here is that if we can keep interferon from being made, we can nip this defense in the bud. The second strategy is to interfere with the interferon-induced activation of antiviral genes. If the interferon block is not completely successful—and usually it isn't—we can try to keep the interferon that is made from triggering the production of proteins that can destroy us. And the final strategy some viruses use to deal with the interferon defense is to blunt the effects of those antiviral proteins. Smart viruses like me employ a "layered defense" in which we encode proteins that disrupt interferon production as well as proteins which help protect against the effects. You just can't be too careful.

I: What features of your lifestyle alert a cell that it has been infected?

FV: I like being an RNA virus, I really do. But one of the drawbacks is that as my genome is being replicated, relatively long stretches of double-stranded RNA are produced, at least transiently. For example, when my vRNA is used as a template to make positive-strand mRNA, stretches of double-stranded RNA exist in the nucleus. I can't help it. How else am I going to copy my negative-strand genome? Unfortunately, in uninfected cells, this kind of RNA doesn't exist—and these clever human cells have set a trap for us. Inside these cells are pattern recognition receptors like TLR3, which can detect double-stranded RNA and activate the interferon system.

Another red flag that I have to be very careful about is the 5′ triphosphate at the end of the single-stranded RNA created by my RNA polymerase. It's quite a fine polymerase otherwise, but it produces this "abnormal" 5′ end. And when this unusual termination is detected by the cellular pattern recognition receptor, RIG-I, the interferon warning system is activated.

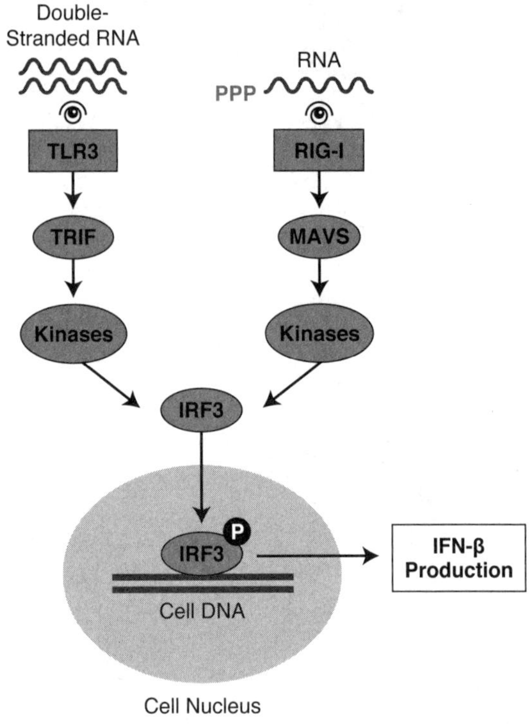

I: And how do you counter this defense?

FV: We influenza viruses encode a protein, NS1, that binds to the double-stranded RNA, masking it so that it cannot be recognized by the pattern recognition receptors. NS1 also binds to RIG-I and inhibits its ability to signal the presence of viral RNA which has a 5′ triphosphate. In addition, once I steal a cap from cellular mRNA and paste it on, my mRNA no longer has a 5′ triphosphate, so cap stealing actually helps me evade detection by the interferon system.

I: Anything else you do to defend against the interferon system?

FV: Yes, there is. The measures I mentioned are helpful because they diminish the amount of interferon synthesized by the cells I infect, but some interferon is still produced. And that's a bummer, because when this interferon is secreted and binds to its receptors on a cell's surface, it induces the expression of many cellular genes with antiviral properties. There are literally hundreds of them! For example, one of these interferon-induced genes encodes a kinase, PKR, which, when activated by double-stranded RNA, can shut down protein synthesis in an infected cell. And that's only part of it. The expression of two other genes, 2′-5′-oligo(A) synthetase and RNAse L, also is induced when interferon binds to its receptors. Double-stranded, viral RNA activates this synthetase, which then activates RNAse L. And when RNAse L is in its active form, it can destroy both cellular and viral RNA. These guys really like to gang up on me! Fortunately, my NS1 protein can cloak my double-stranded, viral RNA, and reduce the activation of these destructive enzymes, allowing protein synthesis to proceed in infected, interferon-warned cells. I'm really proud of my NS1 protein!

I: Yes, I can see how you would be. Any other tricks up your sleeve?

FV: You really want to know everything, don't you! Well, okay, here's another. I'm one of the speediest viruses known to man or woman—influenza viruses don't discriminate on the basis of gender like some other viruses do. We are equally happy to infect men or women! Anyway, within about six hours after I enter a cell, new viruses are produced. Because the NS1 protein is such an effective defense, it usually takes about two days for the interferon system to get totally fired up. And because each cell I infect can turn out new viruses in six hours, a lot of baby viruses can be produced before the interferon system shuts us down. But here's the fun part. During these first two days, my "patient" is contagious, yet feeling no symptoms. Of course, after a couple days, things start to get a little more dangerous—because NS1 can't hold off the interferon system forever. But by then, my job is finished, and, if my patient is fairly social, lots of new hosts will have been infected.

Now, I'd be the first to admit that my efforts to evade the interferon system never are completely effective. But, you see, it really doesn't matter. The tactics I use only have to hold the defenses at bay long enough for me to reproduce. I can do that.

HOW INFLUENZA VIRUSES SPREAD

I: You mentioned that you are a cytolytic virus. Certainly, the "unusual" cell death you cause must activate the immune system. Are you sure that killing the cells you infect is a good idea?

FV: It is risky. That's for sure. But killing cells is essential to my way of life. First, I must reproduce rapidly to survive, and when a virus disrupts cellular protein synthesis by, for example, snatching caps, that cell is going to die. You can count on it. It's an unavoidable consequence of rapid reproduction. But there's an even more important reason for killing cells—and intentionally alerting the immune system.

When cells of the lower part of the respiratory tract are killed, the resulting inflammation triggers the cough reflex which, of course, is designed to clear the airway of foreign invaders like me. However, the result of this cough is that microdroplets containing newly minted viruses are expelled into the surroundings. In fact, a rather heroic cough can deliver virus-containing droplets at about 100 miles per hour, and each droplet can contain millions of virus particles. In addition, because I am a versatile virus, I also can infect and kill cells in the upper part of the airway. In the nasal passages, the immune system, responding to the killing of infected cells, generates chemicals (like histamine) which can stimulate the sensory nerves and trigger the sneeze reflex—also intended to expel invaders. Consequently, by killing cells in both the lower and upper airway, I actually take advantage of the host's natural desire to get rid of invaders, and use the cough and sneeze reflexes to facilitate my spread. In fact, this "double whammy" of cough and sneeze is so effective that in settings in which people live close together (e.g., in a nursing home), I can infect upwards of 80% of potential recipients during a flu epidemic.

I: Sure, but don't you still have a problem with the adaptive immune system? Killer T cells can destroy virus-infected cells and the viruses within them, so aren't you even more frightened of killer T cells than you are of interferon?

FV: Not really. The adaptive immune system takes at least a week to mobilize its weapons. In contrast, newly minted viruses roll off the assembly line in only a few hours. So by the time the adaptive immune system is ready to go, enough new influenza viruses have been made to ensure efficient spread to the next host. With a viral infection, timing is everything.

I: You may be fast, but once the adaptive immune system is at full strength, this defense usually is so vigorous that most of the viruses which remain within the body will be eliminated within a couple of weeks. Doesn't that bother you?

FV: Not at all. Obviously, "clearing" all the viruses from the body is good for my human hosts, because it keeps them from being killed by the infection. But it's also good for me. Members of my family are planning a return trip to visit this host again soon—and as my Uncle Fred used to say, "A dead host is not a good host."

Antigenic Drift

I: Your mention of a "return visit" brings up another point I wanted to ask you about. I know that influenza virus only causes **acute infections**: You are not able to establish a latent or a chronic infection. Therefore, you must be passed to new recipients during the week or so while an infected person is still contagious. Also, the adaptive immune system is eventually activated so that virus-specific killer T cells and antibodies will be produced. Although these weapons may not keep you from completing your initial infection, the people you infect should become "immunized," making a successful return visit impossible. Indeed, we might expect that you would quickly run out of potential human recipients. So how does influenza virus manage to thrive in the human population?

FV: Ah, yes. Who would have thought that a tiny little virus like me could outsmart the whole human adaptive immune system?! You see, I use what you might call a "bait-and-switch" routine to ensure that I always have susceptible humans to infect. Here's how it works.

My RNA-dependent RNA polymerase does not have the capacity to "proofread" its work. As a result, mutations in my genome arise as incorrect nucleotides are inserted into the growing RNA chains. You might think that this would be a problem, but, in fact, it's a great advantage. Because my polymerase makes roughly one mistake in every 10,000 nucleotides it copies, nearly every virus produced in an influenza-infected cell will be a mutant. Consequently, my genetic code "drifts" as mutations are introduced during copying, and some of these mutations can change the shape of the hemagglutinin and neuraminidase proteins enough to prevent binding by neutralizing antibodies. Although anti-hemagglutinin antibodies are the most important ones for us to avoid—because they can block infection—anti-neuraminidase antibodies also are very scary. Those antibodies can bind to the neuraminidase protein and prevent it from functioning efficiently, severely limiting the number of virus particles that escape successfully from infected cells. I've heard horrible stories about viruses that were trapped on the surface of infected cells by those sticky neuraminidase proteins.

I: A terrible fate, I'm sure. Influenza strains are defined by the neutralizing antibodies that can bind to either the hemagglutinin protein or to the neuraminidase protein, and at least 16 hemagglutinin "subtypes" (H1–H16) and nine neuraminidase subtypes (N1–N9) have been identified. Does it annoy you that influenza strains are given letters and numbers such as H1N1 instead of names?

FV: Yes, it is a bit demeaning. Even hurricanes have names, and yet influenza viruses—which certainly affect more people than do hurricanes—get letters and numbers.

I: So if I understand you correctly, your bait-and-switch strategy goes like this: During an infection, the "current" strain of influenza may mutate so that when this "new" virus spreads to individuals who were infected by the original strain, they can no longer be protected by the antibodies they have on hand.

FV: Exactly. I offer the immune system one strain to defend against during the initial infection—and then, when this "bait" has been taken and antibodies have been made, I mutate in order to "switch" to another strain which the immune system has never seen before. Because of this "**antigenic drift**," individuals with "outdated" antibodies, who are immune to the original virus, can be reinfected by the mutant strain. Antigenic drift is responsible for the local outbreaks (epidemics) of influenza infection that occur every year or two. Bait and switch. One of my finest inventions!

Antigenic Shift

I: Any other tricks up your sleeve that allow you to reinfect your human hosts?

FV: Yes, I do have one more, and it's totally sick—no pun intended! Virologists call it "**antigenic shift**," and it's unique to type A influenza viruses like me. Types B and C can't use this ploy, because neither virus has a non-human host which can produce a strain of virus that infects humans. I, on the other hand, am much more versatile. For example, I can infect birds and, in particular, water fowl. In fact, human strains of influenza A virus were originally avian viruses that, by mutation, acquired the ability to infect humans. One of the intriguing features of this "adaptation" is that whereas type A influenza causes an acute respiratory disease in humans, in ducks we infect cells in the digestive tract, and the infection is spread by the fecal-oral route.

Wild ducks gather on lakes prior to beginning their migration. There, infected ducks defecate and newly made viruses are excreted at high concentrations in the feces. These viruses are subsequently ingested by young (previously uninfected) ducks who drink the virus-containing lake water. Because the number of ducks is great, and the infection is so efficient, wild ducks like these represent a large reservoir of novel strains of influenza A virus. It's a lovely story—so beautiful, so natural.

I: But what about your Uncle Harold—the virus who tried the fecal-oral route and didn't live to tell about it? How can influenza virus survive in the acidic conditions of the duck digestive tract and cause an intestinal infection?

FV: That's the beauty of having an error-prone polymerase. Because of mutations introduced by our polymerase, duck versions of influenza A virus are less sensitive to acidic conditions than are human strains of flu. You see, slight changes (due to mutations) in the genetic code of a virus can dramatically alter the virus' route of entry and the resulting pathology.

I: I'm sure that's true, but humans rarely are infected directly by avian influenza A viruses, so why are ducks and birds reservoirs for new human influenza strains?

FV: You're right about the birds. Avian influenza viruses usually reproduce poorly in human cells. However, a pig can be a host to <u>both</u> avian and human influenza A viruses, and occasionally viruses of both origins infect the same pig cell. When this happens, new strains can be produced that are part human and part bird. This antigenic shift is possible because of my segmented genome. Each segment could be contributed either by the human virus or by the bird virus. Most of these hybrid viruses are not infectious and represent an evolutionary dead-end. Sometimes, however, this "mixing in the pig" results in a hybrid virus that can successfully infect humans. Importantly (at least for me), antigenic shift is a mechanism which allows me to make big changes in my genome. Of course, the probability that a reassortment of viral gene segments will take place according to this scenario is greatest in places where large numbers of pigs, ducks, and humans are in close contact—for example, in Asia. I do love China!

INFLUENZA-ASSOCIATED PATHOLOGY

I: That brings me to some questions about the diseases you cause.

FV: Just a minute, please! I want to set the record straight about the diseases. I take full responsibility for the coughing and the sneezing. I really have no choice on that. I rely on those reflexes to spread. But the fever, muscle aches, headaches, and fatigue simply are not my fault. All those symptoms are caused by the immune response, especially that despicable interferon system. Sure, the fact that double-stranded RNA is made when I reproduce does activate the interferon system. But I can't help that. In fact, I do my very best to reduce the amount of interferon given off by infected cells. They just keep cranking it out!

I: Okay, but what about the pneumonia that is sometimes associated with influenza infections? What do you say about that?

FV: Another bad rap. I have to reproduce quickly to avoid host defenses, and I sometimes end up destroying large areas of ciliated epithelial cells. Of course, if the host's defenses would just ease up a little, I could probably figure out how to infect cells more gently. But no! They come after me with a vengeance. So I'm stuck being cytolytic. Besides, the cells I kill are replaced when undamaged epithelial cells proliferate. Of course, in the interval between killing and replacement, the "paddles" of these epithelial cells are stilled. And as a result, infected individuals do become susceptible to superinfection by other pathogens. The fact is, however, most cases of influenza-associated pneumonia are due to superinfecting bacteria which have easy access to the damaged airways. I don't invite those bacteria to come in. They just show up.

I: And isn't the antigenic shift you are so proud of responsible for the pandemics that have wiped out millions of humans? For example, the 1918 "Spanish" flu pandemic killed more than 20 million people worldwide. You can't be too proud of that.

FV: It's true that mixing in a pig does produce pandemics about every decade or two. This usually occurs when the pig virus hybrid includes a bird gene segment that encodes a novel hemagglutinin protein—so novel that none of the neutralizing antibodies produced during earlier flu infections can recognize the bird hemagglutinin. When this happens, we influenza viruses get to make a whole new start. It's very exciting! And if a pandemic kills a few million humans, it's no big deal. After all, there are about seven billion of you.

I: I think it's about time to conclude this interview!

THE INTERVIEWER'S SUMMARY

Influenza virus infects the epithelial cells which line both the upper and lower respiratory tract, reproduces rapidly, and eventually kills the cells it infects. After using enzymes present in the airway mucus to prepare it for entry, the virus binds to sialic acid residues on its target cell, and is taken into the cell via receptor-mediated endocytosis.

The genome of type A influenza virus is composed of eight segments of single-stranded RNA. The virus carries its own RNA polymerase, which it uses to copy its negative-strand genome to make mRNA, and influenza virus replication takes place in the nucleus of the infected cell. As part of its rapid reproduction strategy, influenza virus removes cap structures from cellular mRNA, and pastes them onto its own 5′ ends. This "cap snatching" ploy helps focus the attention of the cell's protein-making machinery on viral mRNAs. When new viral gene segments have been synthesized, they are coated with virus-encoded proteins, and once the requisite number of gene segments has been gathered together, the virus buds from the cell. As they exit, newly minted viruses acquire an envelope made from a portion of the cell's plasma membrane into which viral hemagglutinin and neuraminidase proteins have been inserted.

The double-stranded RNA and the 5′ triphosphate-terminated RNA generated during viral replication are detected by cellular pattern recognition receptors, and this results in the production of interferon. To reduce both the amount of interferon produced and its effects, influenza virus encodes a multifunctional protein, NS1. This protein can interrupt the pathways that signal recognition of influenza RNA. In addition, NS1 reduces the effects of interferon in interferon-warned cells by "hiding" viral double-stranded RNA from view. And cap snatching "covers" the 5′ triphosphate on newly made viral RNA, which otherwise would alert the interferon system. These tactics allow influenza virus to hold off the interferon system just long enough for it to complete its infection.

The death of cells infected by this cytolytic virus is one of the signals which alerts the adaptive immune system, and, as a result, virus-specific B and T cells are produced during an influenza virus infection. However, influenza virus replicates so quickly that by the time this defense is fully mobilized, the virus already has produced copious prodigy, and is on its way to infect other hosts. Not only does this clever virus "outrun" the adaptive immune system, it uses the cell death that alerted this defense to its advantage: The destruction of cells in the airway and the immune response to the infection trigger the sneeze and cough reflexes by which the virus spreads. Indeed, all the symptoms commonly associated with an influenza infection—from a sore throat to pneumonia—are the consequence of the virus' "decision" to reproduce in a style that induces interferon production and destroys cells that line both the upper and lower respiratory tract.

Killer T cells and anti-influenza antibodies eventually destroy any virus that remains in the host, and leaves that victim immune to reinfection by the same influenza virus strain. However, during an infection, the error-prone RNA polymerase of influenza virus makes small changes in the virus' genetic code, and this "antigenic drift" can produce new strains of virus. In addition, the swapping of bird or animal gene segments for human gene segments in type A influenza virus can result in more dramatic changes in the viral genome termed "antigenic shift." Together, these two mechanisms can create novel flu strains which the adaptive immune system of a previously infected human does not recognize—enabling influenza virus to infect the same human more than once.

GENERAL PRINCIPLES

During this interview, we have identified principles which, although they were illustrated by examining the lifestyle of influenza virus, are applicable to many other viruses. These general virological principles are worth remembering.

1. The coat of every virus must be stable enough to protect the viral genome from conditions outside cells, yet once the virus has entered its target cell, the coat must be at least partially disassembled to allow replication of the virus' genetic information.
2. The genomes of single-stranded RNA viruses can either be "positive strand" or "negative strand." Positive-strand viral RNA is defined as RNA which is ready to be translated into proteins — so it is synonymous with viral messenger RNA. Negative-strand RNA is the "opposite" strand, so a complementary copy of it must be made to produce viral messenger RNA.
3. Human cells do not have polymerase enzymes which can replicate RNA molecules, so every virus which has an RNA genome must encode its own polymerase.
4. Enveloped viruses always arrange to have viral proteins inserted into the cellular membranes through which they bud. One or more of these "extra" proteins will form the "plug" that will engage the "socket" (receptor) on the surface of the next cell the virus intends to infect.
5. The host defense which all viruses fear most is the interferon system, and viruses have evolved three ways of protecting themselves from this defense. Viruses can encode proteins which either interfere with their detection, block the production of interferon, or blunt the effects of interferon once it has been produced.
6. The tactics which a virus uses to evade host defenses need not be 100% effective. They simply must hold the defenses at bay long enough for the virus to reproduce.

THOUGHT QUESTIONS

1. Why does influenza virus "choose" to infect the epithelial cells of the human airway?
2. How does influenza virus protect itself from the interferon defense system?
3. Why is cap snatching important for the lifestyle of influenza virus?
4. When influenza virus infects a human host for the first time, how does it defend itself against the weapons of the human adaptive immune system?
5. How does the virus maintain a pool of non-immune human hosts?
6. What are the pros and cons of the virus killing the cells it infects?
7. What features of the virus lifestyle are responsible for the pathological consequences of an influenza infection?
8. What causes influenza epidemics and pandemics, and how do they differ?

Chapter 5

Rhinovirus: A Virus That Surrenders

BACKGROUND

The average American suffers from a rhinovirus infection about once a year, and roughly half of all cases of the "common cold" are caused by this virus. So rhinovirus would make our top-12 list on this basis alone. In addition, the contrasts between the lifestyles of rhinovirus and influenza are truly remarkable. Although these two viruses are transmitted in the same way, they have devised very different strategies to solve their problems, resulting in markedly different pathological outcomes.

A RHINOVIRUS INFECTION

Interviewer: You are a famous respiratory virus, so I assume that you infect people via coughs and sneezes, and that you target the epithelial cells that line the respiratory tract—just like influenza virus.

Rhinovirus: I don't know about the famous part. Most people don't even know my name. I'm not a big shot like influenza virus. But you are right that I use coughs and sneezes to spread. Indeed, a single sneeze droplet from an infected person can contain as many as 100 million rhinoviruses. Nevertheless, there is a subtle difference in the location of the cells influenza and I target. Because of the structure of the protein **capsid** that protects my genetic information, I am most infectious at temperatures that are somewhat below normal body temperature. In fact, my favorite temperature is about 91°F—the temperature found in parts of the nose and upper airway. Consequently, I specialize in infecting cells of the upper respiratory tract. In contrast, influenza has full run of the airway. He is clearly much more versatile—and much more important.

I: How refreshing to interview a virus who is not so full of himself.

RhV: It's hard to be boastful when you're named after an ugly jungle animal with a big nose!

I: I see your point. In addition to the blanket of mucus that protects the airway, you respiratory viruses also must contend with the "paddles" on ciliated epithelial cells which move this mucus.

RhV: Yes, that's true, but I am very fortunate in that respect. The "mucociliary escalator" actually runs in two directions. There is an "up escalator" that carries mucus upward from the lower respiratory tract, and there is a "down escalator" that moves mucus downward from the nasal cavity. The idea, of course, is that invaders in all parts of the respiratory tract will be swept in the direction of the throat to be swallowed or coughed up. Influenza viruses heading toward the lower parts of the airway must "swim" against the upward current of mucus. That's gotta be tough. In contrast, I actually use the down escalator to catch a free ride to just where I want to go—the interior of the nasal cavity where the temperature is just right for optimal infection. There is a danger, however. If I ride the escalator too far, I may be swallowed and subsequently destroyed by acids in the stomach.

I: And when you reach your target cells, how do you enter?

RhV: Depending on the strain of rhinovirus, we can bind to two different molecules on the surface of the cells we infect: either ICAM-1 or the LDL receptor. In either case, binding to the cellular receptor results in the destabilization of my protein capsid, and the release of my genetic material directly into the cell's cytoplasm.

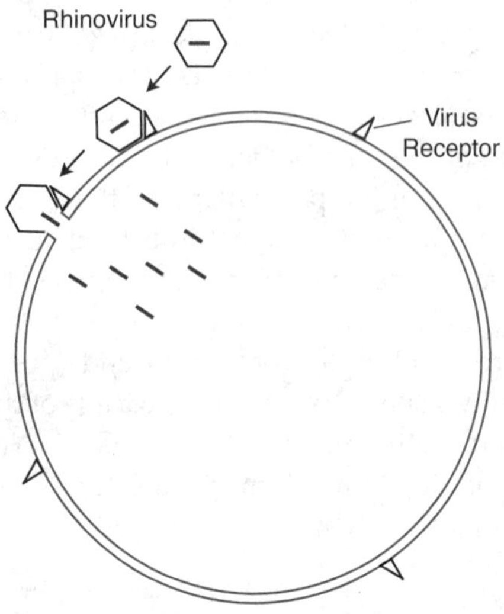

RHINOVIRUS REPRODUCTION

RhV: Like influenza virus, I have a single-stranded RNA genome, but there the similarity ends. The influenza virus genome consists of multiple segments of negative-strand RNA. In contrast, my genome is a single piece of positive-strand RNA. I know I'm biased, but I think it is an advantage to be a positive-strand virus. When negative-strand viruses enter a cell, they must use their polymerase to convert their genome into messenger RNA. In contrast, positive-strand RNA viruses like me come into the cell ready to be translated into protein by the cell's protein-making machinery. This feature increases the speed with which viral proteins can be produced, allowing me to get a jump on host defenses.

I: I can see that being "positive" would be an advantage. Anything special about your replication strategy?

RhV: I think so. When my genomic RNA is translated, the product is one long protein (the "polyprotein") which almost immediately cuts itself into smaller pieces to produce the various viral proteins. One of these proteins, the viral RNA polymerase, makes complementary copies (cRNAs) of my original viral RNA—and then makes many complementary copies of these negative strands to produce the new, positive-strand viral genomes (vRNAs). And all the action associated with rhinovirus replication takes place in the cell's cytoplasm, not in the nucleus. It's quite efficient. One big protein, and off I go!

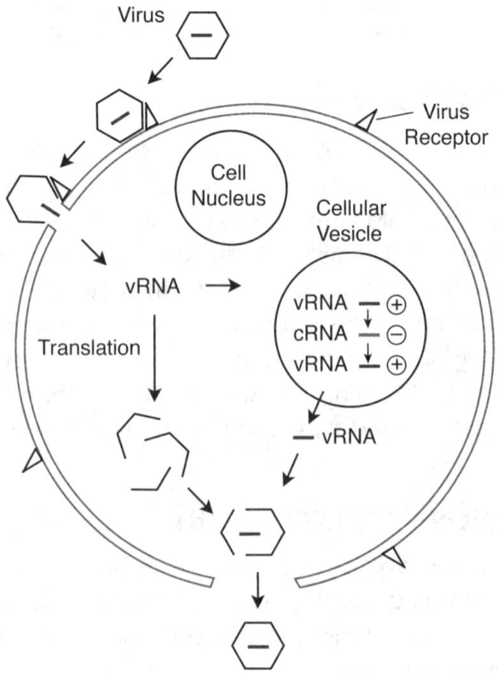

I: Influenza virus uses a "cap stealing" gambit as a way of biasing protein synthesis in favor of the virus. I imagine that a polite virus like yourself wouldn't resort to such a dirty trick.

RhV: Certainly not. Stealing is not my style. I use a different, more honest strategy to take over protein synthesis in the cells I infect. Rhinovirus RNA contains a special initiation sequence, an internal ribosome entry site (IRES), which allows ribosomes to begin translation of viral messenger RNA without the benefit of the cap structure that normally is needed to tell ribosomes where to begin their work. And to take full advantage of this wonderful feature, I encode a protein, 2A(pro), which inactivates two cellular proteins, eIF4GI and eIF4GII, that are required for cap-dependent initiation of cellular mRNA translation. Consequently, I don't need a cap, and the translation of cellular mRNA, which has a cap, is shut down.

I: That does seem more honest—or at least more straightforward. But is that strategy effective?

RhV: Yes, indeed, it is! In fact, I am so adept at taking over the host cell's biosynthetic machinery that in about seven hours, one infected cell will produce thousands of new rhinoviruses. The newly synthesized rhinovirus genomes are enclosed in a single protein shell (capsid) constructed from virus-encoded proteins. No gooey envelopes for us.

I: Impressive. But what about the cells you infect? What happens to them?

RhV: When all those cute little rhinoviruses are ready to go, they burst from the infected cell, leaving it dead or dying.

I: Oh, my! And I thought you were so polite.

EVADING HOST DEFENSES

I: I understand that all viruses must somehow evade the interferon defense. How do you accomplish this?

RhV: Rhinovirus-infected cells produce and **secrete** much less interferon than do influenza virus–infected cells. While cap snatching by influenza virus does bias protein synthesis in favor of viral protein synthesis, my strategy of using an internal initiation site and blocking the initiation of cap-dependent, cellular protein synthesis is much more efficient. In fact, by using these two tricks, I am able to shut down essentially all protein synthesis from capped, cellular RNA in less than two hours. Because the shutdown of cellular protein synthesis is so fast, relatively little interferon (a cellular protein) is produced by the infected cell. And to be doubly careful (as most viruses are!), I disrupt the cellular system used to transport interferon out of rhinovirus-infected cells. Consequently, very little interferon is exported out of (secreted by) the cells I infect to warn nearby cells of an impending attack.

I: So you are careful to limit interferon production. Does this mean that you don't try to protect yourself against the effects of any interferon which might be produced? Most viruses do that, you know.

RhV: Not me. Why should I invest valuable genetic "capital" in defending myself against the effects of interferon when I am so good at blocking interferon production—at least long enough for me to reproduce? In fact, you humans could probably avoid ever getting a rhinovirus infection by sniffing interferon alpha every day—although you probably wouldn't want to live with the side effects. Nasty stuff, that interferon.

I: It is the adaptive immune system (B and T cells) we humans have to thank for defending us against a lethal influenza infection or from a second infection by the same strain of influenza. The adaptive immune system must also pose a problem for you.

RhV: Not really. The immune system's defense against a rhinovirus infection is mediated mainly by the innate immune system (e.g., complement, professional phagocytes, and interferon). I reproduce quickly, and any rhinoviruses that remain in the body after most of us have left "surrender" to the innate immune system—usually before the adaptive system is alerted. This strategy leaves the immune system with a warm, fuzzy feeling of success: The innate system is all aglow because it has saved the life of a human, and the adaptive immune system is delighted that it doesn't even have to get out of bed! It's a win-win scenario.

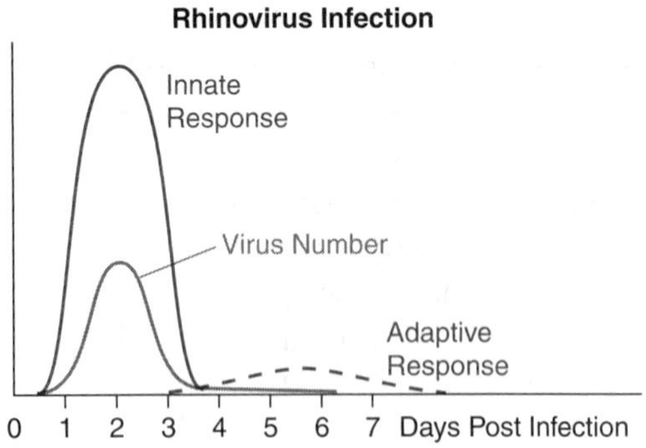

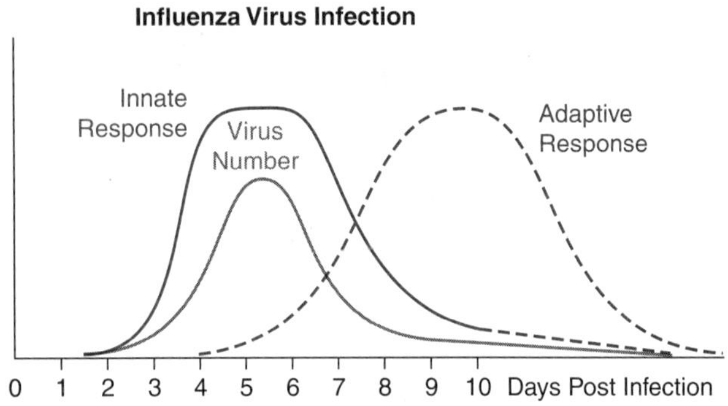

VIRAL SPREAD

I: How do you spread to new hosts?

RhV: Sneezes. The tissues of the upper respiratory tract are richly provided with capillaries that lie very near the surface. These capillaries function as heat exchangers which, by transferring heat from the blood, warm the room-temperature air as it makes its way toward the lungs. However, this extensive capillary system is also a major target for the cytokines and other inflammatory mediators given off during the innate system's defense against a rhinovirus attack. These chemicals cause the capillaries to leak, and the fluids that escape produce the runny nose—which I'm sure you have enjoyed! These same chemicals trigger the sneeze reflex, and off we go.

I: I suppose that, despite your best efforts, the adaptive immune system sometimes is alerted, making your host immune to reinfection. And since you depend on spreading from person to person, it seems like this could be a difficulty. Can you use antigenic shift to make large changes in your genome and avoid this problem?

RhV: Clearly that's impossible. Rhinovirus RNA is all one piece, and no non-human reservoir of rhinovirus genomes exists. So the antigenic shift evasion strategy employed by a segmented virus such as influenza A is not available to us. However, my RNA polymerase is error-prone. So like influenza, I can use antigenic drift as an additional mechanism to stay one step ahead of the adaptive immune system. In fact, as a result of antigenic drift, over 100 different strains of rhinovirus are currently circulating in the human population. And, as I mentioned before, because we surrender to the innate immune defense so soon after infection, the adaptive immune system usually doesn't become fully activated. As a result, neutralizing antibodies generally are not made in sufficient quantities to protect against a subsequent rhinovirus attack—even by the same strain. Consequently, we are able to infect the same hosts over and over again. I am told that in terms of the number of people infected (or reinfected) each year, I am the world's most "popular" virus. It's nice to be so popular!

PATHOLOGICAL CONSEQUENCES OF A RHINOVIRAL INFECTION

I: I'm not sure "popular" is the word I'd use. You certainly aren't all that popular with me when I get my yearly cold with that awful runny nose.

RhV: Sorry about that, but a virus has gotta do what a virus has gotta to.

I: Seems like I've heard that before.

RhV: Besides, I'm a much kinder virus than influenza. Although we both only cause acute infections, influenza engages in a protracted battle with the adaptive immune system. I surrender after just a few days of infection. In fact, you might consider me to be a "pacifist." And because I stick around for the shortest possible time, the symptoms caused by a rhinovirus infection usually are very mild. Indeed, my mantra is: Make 'em just sick enough to sneeze and infect other people—but not so sick that they stay home in bed!

Of course, in some cases, the sneezing—which is essential for my transmission—can spread a rhinovirus infection up into the Eustachion tubes and into the sinuses, causing middle ear infections and sinusitis. But this is the exception, not the rule. And the fever—I'm certainly not in favor of the fever. That is caused by the interleukin-1 produced by macrophages responding to a rhinovirus infection. I don't like it hot because fever makes it harder for me to reproduce efficiently. All of the typical, flu-like symptoms of fever—muscle aches, fatigue, and headaches—are much milder in a rhinovirus infection than in an influenza infection. That's because these symptoms are caused in large part by interferon, and I try my very best to limit the amount of interferon given off by infected cells.

I: And I suppose you don't cause pneumonia?

RhV: Almost never. Although I do kill the cells I target, I usually only infect patches of epithelial cells in the upper airway. This is in contrast to influenza virus, which typically destroys much larger areas of epithelial cells throughout the respiratory tract, causing much more inflammation. Also, because I reproduce best below the core body temperature, rhinovirus infections generally are confined to the upper respiratory tract. I do all this for your benefit, of course. I'm always thinking of the welfare of my host.

I: Sure you are!

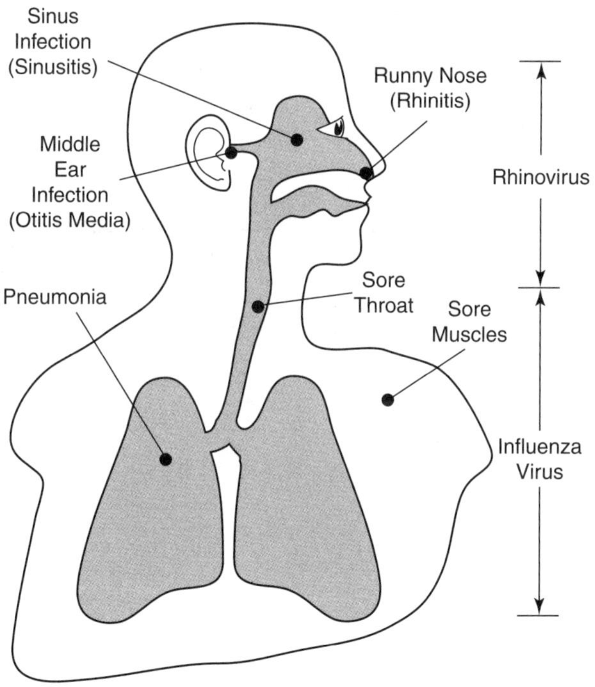

THE INTERVIEWER'S SUMMARY

Rhinovirus invades the respiratory tract in overwhelming numbers, rides the mucociliary escalator down until it reaches the part of the airway which is slightly cooler than core body temperature, penetrates the mucosal barrier, binds to receptors on the underlying epithelial cells, and injects its genetic information directly into the cytoplasm of its target cell. The rhinovirus genome is a single piece of positive-strand RNA that is quickly translated into a long polyprotein which then cleaves itself into smaller proteins.

Viral mRNA contains internal ribosome entry sites, which allow viral RNA to be translated without the cap structure required for the translation of cellular mRNA. This novel feature enables the virus to interfere with the synthesis of cellular proteins, causing the cell's protein synthesis machinery to favor the production of viral proteins. This rapid shutdown of cellular protein synthesis is one of the reasons why cells infected with rhinovirus produce very little interferon.

Newly made viral RNA is clad in a single protein capsid, and is released as the infected cell ruptures. The entire replication process only takes about seven hours. Rhinovirus targets cells in the upper airway, kills relatively few of these cells, and quickly surrenders to the innate system. The result is that neutralizing antibodies frequently are not produced in sufficient numbers to protect against reinfection by the same strain. Moreover, the rhinovirus mutates rapidly, so new strains are continually being produced. Its ability to avoid the adaptive immune system and infect the same hosts over and over again is a major reason rhinovirus is so successful. Because a rhinovirus infection is over so quickly and produces so little interferon, the most common symptoms are just the runny nose and the sneeze—which, of course, helps transmit the virus to the next victim.

GENERAL PRINCIPLES

1. Almost all human viruses with RNA genomes replicate in the cell's cytoplasm, not in the cell's nucleus — influenza virus being a notable exception.
2. All viral RNA polymerases are error-prone, because they cannot proofread their work.
3. Viruses are totally dependent on the cell's translation machinery to synthesize viral proteins.
4. Many viruses alter their host's translation machinery to favor the production of viral proteins over cellular proteins.
5. Every virus activates the interferon defense system, at least to some extent.
6. Viruses which reproduce in a matter of a few hours usually kill the cells they infect.

THOUGHT QUESTIONS

1. What are the major differences in the tactics employed by influenza and rhinovirus to infect, reproduce, evade host defenses, and spread?
2. How do influenza virus and rhinovirus focus host translation machinery on the production of viral proteins?
3. Non-segmented viruses cannot use antigenic shift to avoid the adaptive immune system. Why not?

Chapter 6

Measles: A "Trojan Horse" Virus

BACKGROUND

The final respiratory virus in our Parade is the measles virus. Although measles, influenza, and rhinoviruses all enter the body by inhalation, influenza and rhinovirus infections are contained within the airway. In contrast, measles virus causes an acute, systemic infection of cells throughout the whole body. This illustrates the important point that even viruses which use the same entry portal can cause very different types of infection with very different pathological consequences—because they have evolved different strategies to solve the problems of infection, reproduction, evasion, and spread.

MEASLES VIRUS INFECTION

Interviewer: You are a respiratory virus, so I assume that, like influenza virus and rhinovirus, you are transmitted by the cough or sneeze of an infected individual. However, in contrast to flu and rhinovirus—which within a few days after infection produce large quantities of virus, cause typical cold or flu-like symptoms, and trigger cough and sneeze reflexes—measles virus usually causes no disease symptoms until about 10 days after infection. This suggests that there is something very different about the way you infect your hosts.

Measles Virus: Indeed there is. I use a "Trojan horse" strategy to spread my infection throughout the body—and eventually back to the airway.

I: I didn't know viruses read Greek mythology!

MV: There's probably a lot about viruses you don't know. But let me explain what I mean by a Trojan horse strategy. You see, for many years, virologists have searched for the receptor by which measles virus gains entry into cells of the respiratory epithelium, and only recently have they figured out that I use a cellular protein (CD150) as my receptor when I infect humans. However, this "breakthrough" really confused them because CD150 isn't present where they expected to find it—on epithelial cells in the airway. I love to keep them guessing!

I: But if CD150 isn't found on airway cells, how do you manage to initiate an infection?

MV: Careful! I didn't say CD150 was absent from airway cells—only that it was not on the surface of epithelial cells in the airway. Unlike influenza and rhinovirus, which initiate an infection by entering airway epithelial cells, I infect immune system cells (dendritic cells) which are stationed in the airway to protect against invaders. These cells express the CD150 receptor protein on their surface.

I: But why would you want to infect dendritic cells? That's very strange for a respiratory virus.

MV: You may think it strange, but it is all part of my Grand Plan. Dendritic cells act as sentinels which are on the lookout for viruses and other attackers. When these cells detect an invasion, they travel to nearby lymph nodes, and alert the adaptive immune system that an attack has occurred. By binding to CD150 on the surface of these cells, I can enter these traveling cells before they begin their journey.

I: I'm confused. Why, when you are a respiratory virus, would you want to infect dendritic cells—which are leaving the airway and heading for a lymph node? Lymph nodes don't sneeze, after all, so what's the deal?

MV: Very funny! But you're right. Who could imagine that a respiratory virus would want to hitch a ride with a dendritic cell and end up in a lymph node? But that's the genius of my Grand Plan. It's totally unexpected.

I: Interesting, but please go on.

MV: An important feature of my strategy is that not only do dendritic cells display the CD150 receptor protein, but my reproductive style is just right to infect these cells "productively"—so that they synthesize large quantities of virus. You may know that it is not enough for a virus to bind to and enter a cell. To reproduce efficiently, the reproductive style of the virus must be compatible with the materials and the biosynthetic machinery available within that particular cell type. For example, influenza virus also can enter dendritic cells, but the result is an **"abortive" infection** in which little or no virus is produced—so my Grand Plan won't work for influenza virus.

Once I reach a lymph node, the viruses produced by the traveling dendritic cell infect other immune system cells in the lymph nodes (e.g., macrophages), turning these nodes into virus factories. The resulting viruses are then transported via the lymph to the bloodstream, allowing the infection to be spread throughout the body.

I: So, if I've got this right, measles virus infects dendritic cells in the airway, uses these cells as Trojan horses to carry the virus into draining lymph nodes, turns these nodes into virus factories, and establishes a systemic infection. But why would you want to cause a system-wide infection?

MV: Now we get to the good part. Because many immune system cells are gathered in lymph nodes, infection of these cells produces a whole hoard of new viruses. The bloodstream will be filled with them. Importantly, these viruses can escape the bloodstream into the tissues, including the tissues that surround the respiratory tract. Now I'm going to tell you something few people know, so I want you to keep it quiet. This conversation is covered by "virus-interviewer privilege," isn't it?

I: Yes, of course. My lips are sealed.

MV: Okay. In addition to CD150, there is another receptor I can use, nectin-4, which is expressed on airway epithelial cells—but only on the surface of these cells that faces away from the airway, the basolateral surface. You see, these epithelial cells are "polarized," so the receptors on the basolateral surface are different from those on the surface which faces the airway, the apical surface. The result is that although I can't infect via the apical surface of airway cells (because that surface lacks receptors), I can infect the basolateral surface—which is just exactly where I end up by creating a systemic infection! And because the infection is systemic, I don't infect just a few cells in the airway—as usually happens during the initial infection by a respiratory virus. Indeed, my Grand Plan is so successful that there will be measles viruses crawling all over the basolateral surface of epithelial cells up and down the respiratory tract. What fun!

I: Sounds rather creepy. But after you have infected the epithelial cells from their "back side," what happens next?

MV: Here's the payoff: When new measles viruses are produced in polarized epithelial cells, they are released preferentially from the apical surface of these cells—right into the airway where they can be spread by a cough or a sneeze!

I: That is a Grand Plan! Other respiratory viruses could learn a thing or two from you.

MV: Yes, they could.

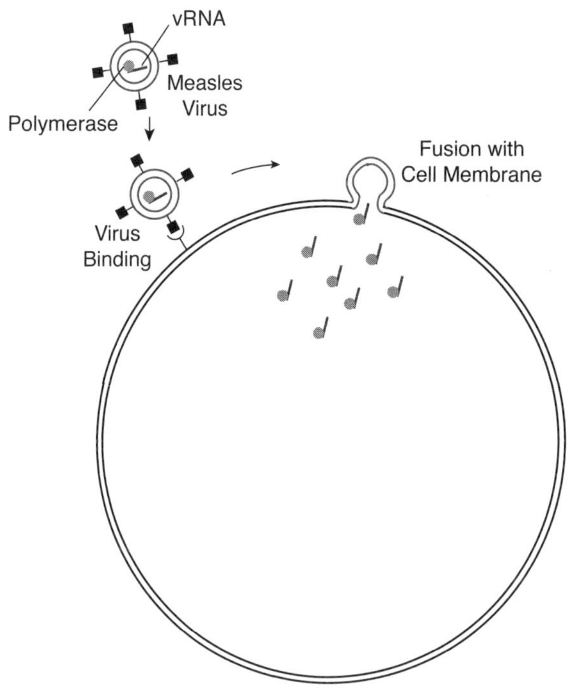

MEASLES VIRUS REPRODUCTION

I: You are a negative-strand RNA virus like influenza virus, so I presume you have a similar strategy for reproduction.

MV: Not really. Although measles virus resembles influenza virus "structurally," in that the measles genome is coated with virus-encoded proteins, and enclosed in a cell-derived envelope, we reproduce very differently. The reason is that the measles genome is not segmented—it is a single piece of RNA. Moreover, I employ an entry strategy that has similarities with that used by rhinovirus, not influenza virus. When my viral hemagglutinin protein binds to its cellular receptor, another of my viral envelope proteins (protein F) promotes the fusion of my envelope with the cell's plasma membrane. When this happens, my RNA and the proteins that coat it are released into the cytoplasm, where replication takes place.

MV: My personal RNA polymerase remains associated with my genome, and once in the cytoplasm, I use this enzyme to make positive-strand copies. However, instead of making one long transcript, the viral polymerase stops at specific sites along the RNA, releases the RNA that has been created, and then restarts—producing a total of six short, positive-strand mRNAs. These are then translated by cellular ribosomes to produce the various measles proteins.

I: That seems like a clever way to get many proteins from one piece of mRNA. But don't you then have a problem? How do you manage to make more full-length, negative-strand RNA molecules to use as genomes for new viruses? Producing complementary copies of the six mRNAs would yield strands of the correct polarity, but these short pieces of RNA would somehow have to be "stitched together" to make your long, negative-strand genome.

MV: Yes, I suppose I could have evolved a way of sewing short pieces of RNA together, but instead, I came up with a much tidier, and, I must say, more elegant way to produce my genomic RNA. One of the proteins I encode, the N protein, can bind to the original, full-length, negative-strand RNA and mask the signals that formerly instructed the polymerase to stop and restart. At the beginning of an infection, there are very few N proteins, so the polymerase usually makes short pieces of mRNA. However, by the time I'm ready to start packaging new genomes, a large number of N proteins have been synthesized—enough to mask the polymerase stop sites. As a result of this masking, the polymerase produces a full-length, positive-strand RNA molecule—which can then be copied many times to make new, negative-strand viral genomes. All this action takes place in the cytoplasm of the infected cell, and once the freshly made genomes have been coated with virus-encoded proteins, they bud from the cell surface, picking up parts of the cell membrane as an envelope. Of course, my envelopes contain, in addition to the usual cell-surface proteins, the hemagglutinin and fusion proteins. Otherwise, how would I enter my next host cell? Here's an electron micrograph taken by a clever virologist who caught me in the act of budding from an infected cell.

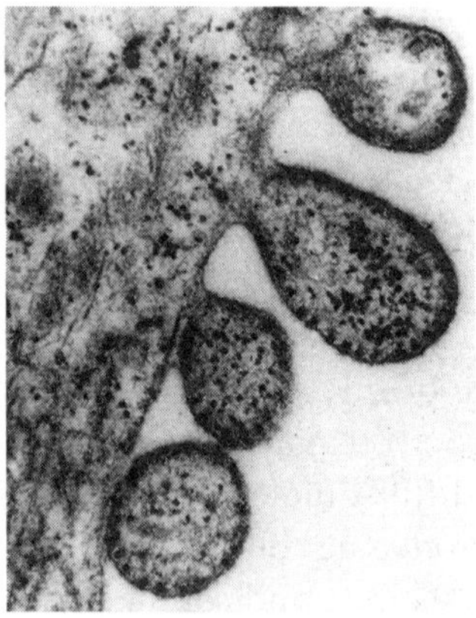

EVADING HOST DEFENSES

I: Your replication scheme generates double-stranded RNA molecules, so I would expect interferon to be produced during an infection. However, I know that when measles virus was isolated from patients and tested, it was found that virus-infected cells synthesize very little interferon. So how do you deal with the interferon defense?

MV: I interfere with the production of interferon, and then with the effects. I think most viruses do it that way. First, I interfere with the phosphorylation of IRF3, so that IFN-β production isn't turned on when TLR3 senses my double-stranded RNA. Virologists still haven't discovered how I manage to block IRF3 phosphorylation. But, of course, I don't care in the least that they haven't figured this out. I just care that it works. And it works really well!

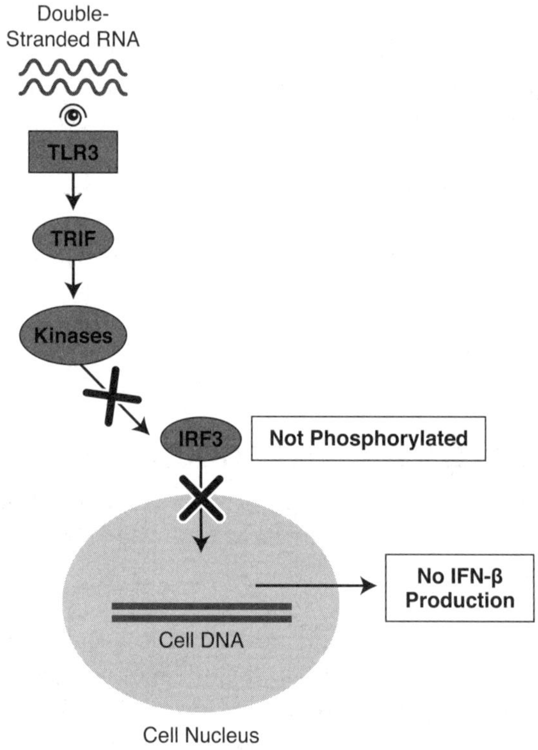

In addition to antagonizing interferon production, I use one of my proteins (V) to interfere with the signal from the interferon receptor to the nucleus of the cell. This protein binds to two of the cellular proteins (STAT1 and STAT2) which make up the ISGF3 complex, and prevents this complex from entering the nucleus and turning on the expression of interferon stimulated genes (ISGs). Consequently, even when some interferon is produced by an infected cell, I can protect myself from its effects by disrupting the signaling pathway which leads to the synthesis of antiviral proteins.

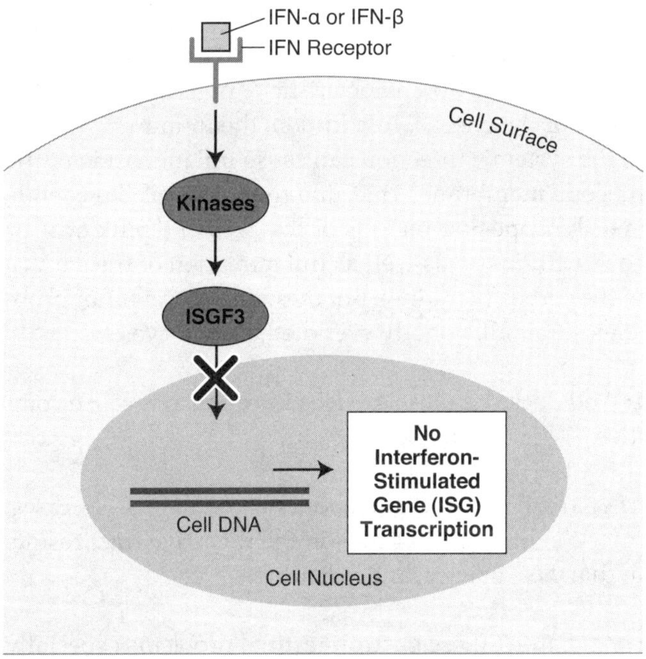

I: It looks like you are quite skillful in dealing with the interferon defense. But what about the adaptive immune system. Your Grand Plan calls for you to establish a systemic infection before you return to the airway to be spread by a cough or sneeze to new hosts. Therefore, I would think that there would be plenty of time for the adaptive immune system to be activated. Of course, your "personal" polymerase is error-prone, so I suppose you might use a rapid mutation strategy to stay one step ahead of virus-specific, neutralizing antibodies.

MV: I wish! Although my RNA polymerase is error-prone, the parts of my envelope that are targets for neutralizing antibodies—the hemagglutinin and fusion proteins—cannot mutate without loss of viral function. As a result, mutated viruses which might escape being recognized by neutralizing antibodies are not infectious, and only one strain of measles virus exists. It's really a shame, but I've learned to live with it.

I: Your answer suggests that you must have some other tricks up your sleeve to avoid the antibody defense.

MV: Of course I do. I can't reveal all my tricks, but here a couple of examples. The fusion protein on the surface of my envelope functions during entry to "glue" my envelope to the outer membrane of the target cell. However, this fusion protein also can cause measles-infected cells to fuse with uninfected, neighboring cells. When this happens, giant cells with multiple nuclei are formed, allowing me to spread "internally" from cell to cell without ever being exposed to the antibody defense. This defense against antiviral antibodies isn't unique, but it is very effective.

I: Since most people don't die from a measles infection, and since you have ways of avoiding the antibody defense, I suspect that the resolution of a measles infection is critically dependent on killer T cells—which can destroy those giant infected cells.

MV: Right you are. Those killer T cells really can do a job on a virus-infected cell. However, you will remember that part of my Grand Plan is to infect immune system cells such as dendritic cells and macrophages. I use these immune cells for reproduction and for transport. But after I have had my way with them, they either are destroyed by the infection, or are rendered incapable of performing their duties as defenders. This results in a dramatic weakening of the immune system of the humans I infect. By suppressing the immune system of infected individuals, I buy time to return to the airway, so that newly made measles viruses can be spread to other humans.

MEASLES VIRUS TRANSMISSION

I: Because there is only one strain of measles virus, people who are infected will be immune to subsequent measles attacks. In addition, humans are the only natural host for measles virus. Doesn't this make it difficult for you to find enough nonimmune hosts to survive?

MV: It certainly does. Measles virus must be spread in an unbroken chain in which every new human target has never been exposed before. Even worse—at least in terms of "viral longevity"—measles virus is so contagious that most humans will be infected as young children. As a result, a relatively large number of adults (probably at least 500,000) must live in close proximity in order to produce enough children to keep me and my relatives in business. Populations of this magnitude did not exist earlier than about 6,000 years ago, so I'm relatively young as viruses go.

I: And aren't vaccinations quite effective in protecting against the single measles strain?

MV: Vaccination! I hate that word! In the United States, vaccination has reduced the number of measles infections to about 300 per year—down from roughly 500,000 per year before vaccines became available. It's almost not worth trying to infect Americans anymore! Fortunately, measles vaccinations are not readily available in underdeveloped countries, so that's where we focus most of our activities these days.

THE PATHOLOGICAL CONSEQUENCES OF A MEASLES INFECTION

I: It would seem that although establishing a systemic infection may be part of your Grand Plan, it probably isn't all that grand for the person who is infected.

MV: It's true that my plan includes a systemic infection that eventually brings me and my "colleagues" back to the respiratory tract in much greater numbers. I am a cytolytic virus, and the killing of virus-infected cells there does trigger a robust inflammatory immune response. As a result, this "second hit" on the airway causes symptoms which resemble those associated with a rhinovirus infection: a little fever and a runny nose. Not a big price for a human to pay for a full-on viral infection, I'd say. Moreover, muscle aches and other typical flu-like symptoms usually are absent—because I'm so careful not to induce the production of large quantities of interferon.

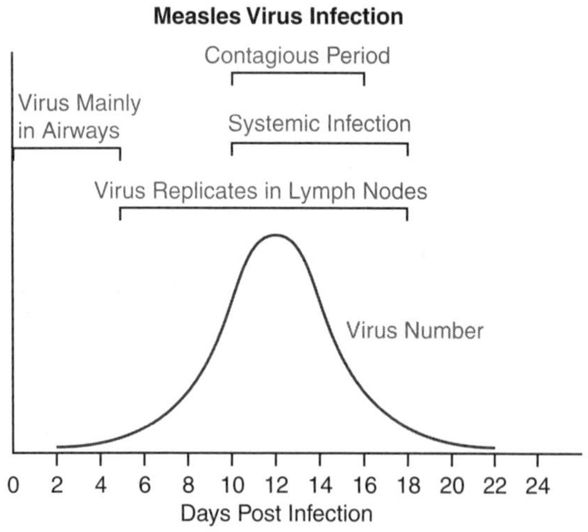

I: But what about the rash?

MV: Oh right, the rash. Okay, I admit that a systemic measles infection can have some pathological consequences that usually do not arise during a rhinovirus or influenza virus infection. T cells responding to the measles attack on skin cells produce cytokines which cause a rash that is one hallmark of a measles infection. Also, infection of the epithelial cells that line the mouth produce the characteristic "Koplik's spots" on the inside of the cheeks. In addition, the immune response to the systemic infection can cause inflammation of the mucous membranes that line the inside of the eyelids and the apposing regions of the eyeball ("pink eye" or conjunctivitis)—as well as inflammation of the surface of the cornea (keratitis). But these really are minor problems. After all, nobody ever died of pink eye.

I: And doesn't a measles virus infection also cause vomiting? What's that about?

MV: Yes, it's true that in about half of all measles cases, infection of epithelial cells in the digestive tract results in diarrhea, nausea, and vomiting.

I: And aren't these gastrointestinal problems especially serious in developing countries where they can exacerbate malnutrition?

MV: Look, it really isn't my intention to cause these problems! I'm a respiratory virus after all.

I: That may be true, but it certainly is your plan to suppress the immune system, which can make a person susceptible to secondary infections. These infections are the main reason why so many children who contract measles die each year in the Third World. In fact, measles infections kill roughly seven million people each year worldwide. Moreover, in about 0.1% of measles cases, the virus spreads to the brain, and about 15% of the time this causes a fatal demyelinating disease (acute postinfectious measles encephalomyelitis). This condition results when the inflammatory response to the virus infection destroys the myelin "insulation" that normally insures rapid transmission of electrical impulses in the brain. And in about one in every 300,000 measles cases, the virus persists in the brains of infected individuals for years, eventually causing a deadly brain disorder, subacute sclerosing panencephalitis. Finally, patients undergoing immunosuppressive chemotherapy or patients whose immune systems are ravaged by the AIDS virus have a mortality rate from a measles infection which exceeds 40%.

MV: Oh, my, I didn't know all that! How sad. I must be more careful. After all, those serious complications clearly are the unintended consequences of the systemic infection—which is an essential part of my Grand Plan. In fact, by the time most of these complications occur, we viruses are long gone, having already spread to a new host. We certainly don't mean to hurt all those children. We love children! Why, without young children to infect, there would be no measles virus.

I: That might be a good thing.

THE INTERVIEWER'S SUMMARY

Although measles virus is spread by the respiratory route, its strategy of infection is very different from that of influenza and rhinovirus. Measles virus "bypasses" the epithelial cells that line the airway, and infects the sentinel dendritic cells which are intended to alert the adaptive immune system that there has been an invasion. The virus then takes advantage of the normal workings of the immune system, and uses these dendritic cells as Trojan horses to carry it into nearby lymph nodes. There it infects other immune system cells, travels through the lymphatic and blood systems, and establishes an acute, systemic infection. By "going systemic," the virus is able to produce many new viruses and return in much greater numbers to infect the epithelial cells of the airway—the cells it formerly shunned. Infection of these cells then triggers the cough and sneeze reflexes which transmit the virus to its next host. This strategy of infection helps make measles one of the world's most contagious viruses.

Measles is a negative-strand, RNA virus, but unlike influenza, the measles genome is not segmented. It is a single piece of RNA, which is injected into the cytoplasm of the cells this virus infects. Early in an infection, the measles polymerase uses the genomic RNA to create six different mRNA molecules by a "stop and restart" mechanism. Then, late in infection, these stop sites are masked by a viral protein, and full-length templates are synthesized to be copied to make new viral genomes. These genomes are then cloaked in virus-encoded proteins, and are wrapped in an envelope as the viruses bud from the infected cell.

Measles virus has multiple ways of dealing with the interferon defense: The virus blocks the production of interferon, and then protects itself from the effects of any interferon which might be made as a result of the "leakiness" of this block. Measles virus kills the cells it infects, and its circuitous route of infection gives the adaptive immune system plenty of time to be activated. It is this defense, led by killer T cells, which eventually rids the body of the virus. However, because immune system cells are destroyed or disabled during the viral infection, the immune response is suppressed. This suppression of the adaptive immune system is one way the virus buys time to complete its infection. In addition, measles-infected cells can infect neighboring cells by fusing with them, avoiding antiviral antibodies that normally would intercept viruses as they travel from cell to cell.

Although the virus has an error-prone polymerase, its envelope proteins are so sensitive to structural changes that mutations which would produce new strains are not possible. Infection with the single measles virus strain results in immunity, and humans are the only natural host for this virus. Consequently, a large population is required to supply enough susceptible humans to keep measles virus from dying out.

GENERAL PRINCIPLES

1. Even viruses which use the same entry portal can cause very different types of infection with very different pathological consequences — because they have evolved different strategies to solve their problems of infection, reproduction, evasion of host defenses, and spread.
2. Viruses have two general strategies for entering cells and uncoating. In the first, the virus binds to its receptor on the surface of the cell, "checks its overcoat at the door," and injects its partially uncoated genome into the cell's cytoplasm. The second strategy is called "receptor-mediated endocytosis." During this process, the virus binds to receptors on the cell surface, is completely enclosed in a portion of the cell's plasma membrane (called an endosome), and then is taken into the cell (endocytosed).
3. It is not enough for a virus to bind to and enter a cell. To reproduce efficiently, the reproductive style of the virus must be compatible with the environment and biosynthetic machinery available within that particular cell type.
4. Every virus encodes at least one protein required for replicating its genome.

THOUGHT QUESTIONS

1. Although measles, flu, and rhinovirus all are RNA viruses which attack cells in the respiratory tract, these three viruses have evolved very different ways of reproducing. Discuss the similarities and differences.
2. Why do you think measles virus evolved its strategy of masking internal polymerase stop sites instead of evolving a way to paste mRNA molecules together to produce a template for the production of genomic RNA?
3. Influenza and measles viruses both can infect dendritic cells, but the outcome of the infection is very different. Why is this important for the lifestyle of each virus — and for the host?

Review 1

A Comparison of Respiratory Viruses

In the previous chapters, we discussed three viruses that enter the body via the respiratory tract: influenza, rhinovirus, and measles virus. These respiratory viruses cause only short-term, "acute" infections. Each virus has a single-stranded RNA genome, and each uses its own RNA polymerase to make both viral messenger RNA and new viral genomes. Nevertheless, the genome structures of these viruses are about as different as any three viruses could be: Influenza has a negative-strand, segmented genome; rhinovirus has a positive-strand genome, which is translated to yield a large polyprotein; and the genome of the measles virus is a single piece of negative-strand RNA. Because influenza and measles virus have negative-sense genomes, they must carry their polymerase molecules with them into the cells they infect. Rhinovirus has a positive-sense genome, so it can arrange to have its polymerase synthesized by the infected cell's protein-making machinery.

Influenza virus enters its target cells by receptor-mediated endocytosis during which the virus is taken into the cell enclosed in a cell-supplied endosome. Remarkably, influenza viral replication takes place in the nucleus of the infected cell. In contrast, rhinovirus and measles virus remove their "overcoats" at the surface of the cell, and enter the cytoplasm, where their genomes are replicated. The rhinovirus genome is protected by a protein capsid which is assembled within the infected cell. The influenza and measles virus genomes associate with virus-encoded proteins and then pick up an envelope made from patches of cell membrane.

Influenza virus can infect cells deep in the airway. In contrast, the rhinovirus capsid is relatively unstable at this core body temperature, so rhinovirus preferentially infects the upper regions of the respiratory tract where it is cooler. When influenza and rhinoviruses infect the epithelial cells that line the airway, they reproduce in these cells, and exit directly into the airway. In contrast, measles virus infects dendritic cells which are keeping watch over the respiratory tract, and uses a "Trojan horse" strategy to escape the airway and travel to nearby lymph nodes. From these nodes the virus launches a systemic infection in which endothelial and epithelial cells throughout the body can be infected. This systemic infection brings the virus back to the airway in large numbers so that it can infect epithelial cells from the "back side." Newly minted viruses then exit infected epithelial cells from the "front side," and take advantage of coughs and sneezes to spread to new hosts.

All three viruses replicate through a double-stranded RNA intermediate—which can be detected by cellular sensors that trigger interferon production. Influenza and measles viruses counter this defense by disrupting the signaling pathway which normally would result in IFN-β synthesis. In addition, both viruses have evolved ways to lessen the antiviral effects of any interferon which is synthesized. In contrast, rhinovirus shuts down host protein synthesis so effectively, and reproduces so quickly that relatively little interferon is produced before the virus is on its way to infect a new victim.

Measles virus immunosuppresses its hosts to weaken the immune response, and this virus also can evade antibody surveillance by spreading "internally" when infected cells fuse with uninfected cells. These ploys buy enough time for measles virus to reproduce and infect new hosts. Eventually, however, the adaptive immune system does succeed in destroying any measles viruses which remain, and renders the original host immune to reinfection. Rhinovirus reproduces quickly, and coughing and sneezing spread the virus efficiently to new hosts—usually before the adaptive immune system can be fully activated. This tactic of "reproduce and surrender" is so effective that a person often can be reinfected at a later time by the same strain of rhinovirus. In addition to the trick of surrendering before immunological memory has fully matured, rhinovirus also can use its error-prone polymerase to create new strains, which can elude neutralizing antibodies during a second visit. In contrast, the region of the measles virus' envelope that is targeted by neutralizing antibodies has a complicated, interlocking structure which cannot "drift" to evade these antibodies without loss of function. As a result, there is only one strain of measles virus, and this strain must be transmitted in an unbroken chain in which each new infectee has never been infected before.

Influenza virus neither surrenders (as rhinovirus does) nor escapes from the airway (as measles virus does). Rather, it "stands and fights" in the respiratory tract until it finally is subdued by a potent adaptive immune response. However, to evade immunological memory and to expand the pool of infectable humans, influenza virus uses two "bait-and-switch" strategies. During replication, the error-prone viral polymerase introduces mutations into the influenza genome. The result of this "antigenic drift" is that almost every flu virus is genetically different from every other one. Some of these mutations change the viral hemagglutinin protein, so that neutralizing antibodies, which could bind to the original virus and prevent it from reproducing, now become totally useless in preventing reinfection by the mutant virus. When one of these "escape" mutants enters the population, the result can be an influenza epidemic—a local outbreak which occurs every year or two.

To further expand its list of potential infectees, influenza A virus adds an additional twist to the bait-and-switch routine. The influenza virus genome is made up of multiple RNA segments, and because influenza A virus also can reproduce in birds and pigs, RNA segments from birds or pigs can be picked up by the human type A influenza virus. Sometimes these non-human sequences encode hemagglutinin molecules which no humans have ever seen before, and against which they have no protective antibodies. Consequently, this "antigenic shift," which is produced when bird or pig RNA segments are acquired, can lead to devastating, worldwide influenza pandemics.

A rhinovirus infection typically lasts for only a few days before the innate immune system subdues the infection. As a result, the symptoms of a rhinovirus infection are quite mild—mostly a runny nose and sneezes—which result from rhinovirus' preference for infecting only the upper respiratory tract. In contrast, it is the adaptive immune system which "mops up" after an influenza virus infection. Consequently, in addition to a cough and runny nose, the typical flu symptoms—fever, muscle aches, fatigue—generally last for more than a week. Moreover, because influenza can infect both upper and lower parts of the airway, an influenza infection also can lead to pneumonia. Measles virus infections also are terminated by the adaptive immune system, frequently more than two weeks after the initial infection. The characteristic measles symptoms—a rash, Koplik's spots, and conjunctivitis—all are the result of the virus' "lifestyle choice" to cause a systemic infection.

Chapter 7

VIRUSES WE EAT

Rotavirus: An Undercover Virus

BACKGROUND

If the respiratory route of infection represents the easy way in, the digestive tract is most certainly the hard way. The goal of viruses that use this route (the enteric viruses) is to infect epithelial cells that line the walls of the intestine. To reach these cells, viruses must be able to resist the antiviral defenses present in saliva, survive exposure to the acid pH and digestive enzymes in the stomach, and escape destruction by the enzymes and detergents that bathe the cells in the intestine which these viruses seek to infect. Only a few viruses can do all this, and rotavirus is one of them.

In infants and young children, rotavirus is the major cause of severe inflammation of the intestine (gastroenteritis). Indeed, about one-third of the cases of diarrhea severe enough to result in the hospitalization of young children are caused by rotavirus infections. Worldwide, this virus causes about 500,000 deaths each year, mostly in underdeveloped countries.

If you classify rotaviruses by the different antibodies that bind to them, there are actually seven different "groups" of rotaviruses. However, group A includes the ones that cause most rotavirus-associated disease in humans, so we'll limit our discussion to this one group.

A ROTAVIRUS INFECTION

Interviewer: The people who named you "rotavirus" certainly got it right. *Rota* is the Latin word for wheel, and that's just what you look like: a wheel with spokes.

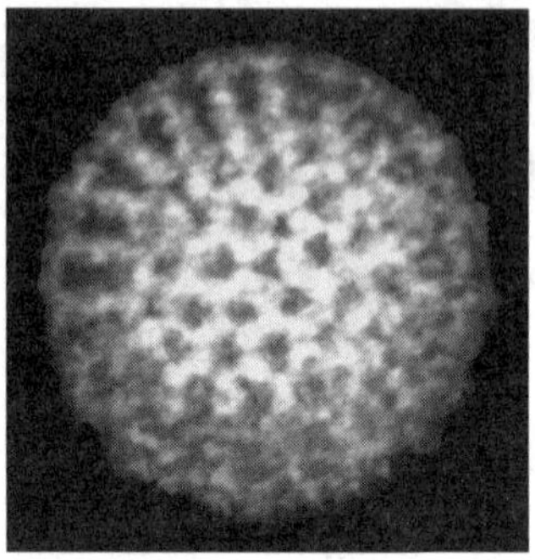

Rotavirus: Not the best picture of me I've ever seen, but it does make the point. What you can't see in that micrograph is that my genome is protected by not one, not two, but three concentric protein shells (capsids). That's really special. Because of my three protective coats, I am able to thrive in the harsh environment of the digestive tract.

My outer capsid includes two different proteins, VP4 and VP7. The VP7 proteins are the primary building blocks of my outer shell, and the VP4 proteins form the spikes.

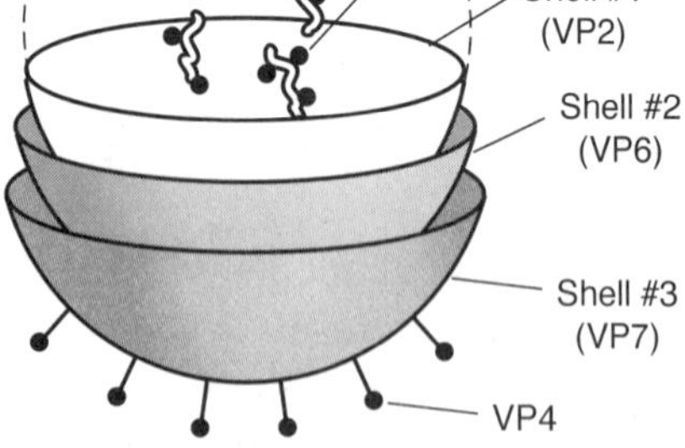

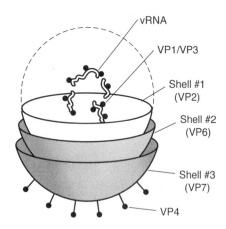

My favorite targets for infection are the columnar epithelial cells at the tips of the villi that line the intestine. Quite tasty, those cells.

I: Yes, I've seen pictures of what those "tasty" cells look like after you've infected them. They look like they have been chewed off!

RV: Yes, I'm a cytolytic virus, so that sort of thing just happens.

I: Whatever. And how do you enter those epithelial cells?

RV: As you might expect, both VP4 and VP7 play important roles in this process. The VP4 protein acts as a plug which engages a receptor molecule on the surface of my target cell. To demonstrate my versatility, I use several different receptors, depending on the virus strain. Once I've "plugged in," I'm taken into the cell by receptor-mediated endocytosis.

VP4 also is involved in helping me out of my "overcoat"—my outer protein shell. However, before VP4 can work its magic, it must be cut by a protease to produce two smaller proteins. What's cool (and important!) is that the villous epithelial cells that line the intestine are bathed in proteases. The "day job" of those proteases is to help cut proteins in the food you eat down to size, so that they can be taken up by your body. However, one of these enzymes, trypsin, is the very enzyme required to cut my VP4. So a rotavirus infection actually uses a digestive enzyme that would destroy most other viruses which dared to enter the intestine! In effect, I use what is normally a barrier defense—the proteases present in the intestine—to prepare me for entry into my target cells.

I: I suppose that the requirement for intestinal enzymes to facilitate the unbuttoning of your overcoat helps explain why you usually don't establish infections in other parts of the body—areas where those enzymes are lacking.

RV: Right you are. Why be greedy? I enter by the most difficult route and leave the rest to other, less studly viruses.

VIRAL REPRODUCTION

I: So at this point, you have removed your outer capsid, but you have two more to go. What's the next step?

RV: My genome is composed of 11 segments of double-stranded RNA which encode 13 proteins. It might seem that the easiest thing for me to do would be to just go ahead and shed my other two coats, and use my RNA polymerase to copy each strand of my double-stranded RNA. In this way, I could produce both the single-stranded viral mRNA needed to encode my proteins, and the double-stranded RNA required for new genomes. That would be the easy way—but it wouldn't be the smart way.

I: Oh? Why is that?

RV: Let's just say that I have my reasons. In any case, I like to replicate "under cover." I use my RNA polymerase (which is packaged inside the virus particle) to transcribe my viral mRNA while my genome is still within the protective environment of its double capsid. The single strands of viral mRNA produced are then spit out into the cytoplasm through holes in my double capsid—something like this:

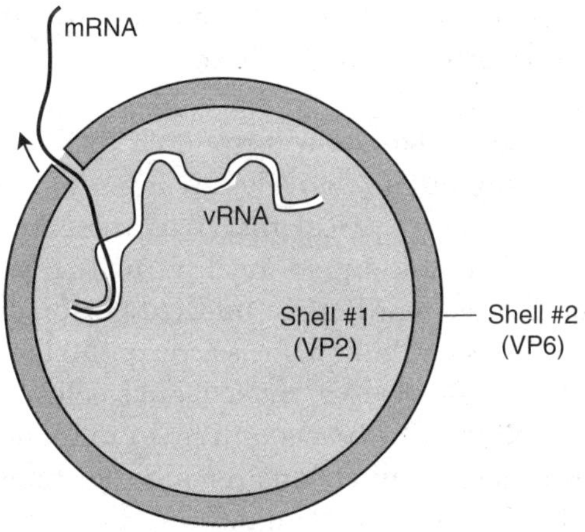

I like to keep my coats on—because it's warm and cuddly in there.

I: I imagine it would be, but how do you produce new viruses?

RV: Again, this process takes place in a protected environment. When it comes time to package new viral genomes, the 11 segments of single-stranded viral mRNA needed for a complete genome are rounded up and cloaked with the proteins that will form the inner coat of the new virus. When this first capsid is in place, my polymerase makes complementary copies of each of the gene segments to yield a complete, double-stranded RNA viral genome. After this process is finished, two more protein coats are added, and all of the newly made rotaviruses exit the infected cell.

I: Some viruses reproduce quickly, and others take their time. How about you?

RV: I'm one of the quick ones.

I: Any tricks to help make this happen?

RV: Yes indeed. You don't get fast without tricks! Cellular mRNA has a poly(A) tail, and for this mRNA to be translated, a cellular protein, PAB1P, must bind that tail and "bend" the RNA molecule so that the 3′ and 5′ ends are brought close together. It does this by "plugging into" a particular site on another cellular protein, eIF4G, which is in a protein complex near the 5′ end of the message. And here's the trick: Rotavirus mRNA is not polyadenylated, but one of my proteins, NSP3, substitutes for PAB1P by binding the 3′ end of my mRNA, bending the mRNA molecule, and plugging into eIF4G. You might think that this would allow both cellular and viral mRNA to be translated. But no! My NSP3 protein binds so much more tightly to eIF4G than does the cellular protein PAB1P that once viral protein synthesis is underway, the binding site on eIF4G usually is occupied by NSP3. The net result is that my viral protein, NSP3, functions to interfere with the synthesis of cellular proteins, and to focus the infected cell's protein-making machinery on producing the viral proteins I need for rapid reproduction.

EVADING HOST DEFENSES

I: You say you like to replicate under cover because it's a "warm and cuddly" environment. But I'm betting you have an ulterior motive for reproducing that way. What's the real reason?

RV: Okay, I see I can't fool you with the warm and cuddly stuff. The real reason I replicate under cover is to evade host defenses. First of all, pattern recognition receptors like TLR3 are on the lookout for double-stranded RNA, so my "replication under cover" ploy helps me evade the host's interferon defense, buying time to reproduce and spread. Also, I have my own capping enzyme which works right along with my personal polymerase. Immediate addition of this cap to my RNA avoids the tell-tale 5′ triphosphate which would alert the RIG-I surveillance protein. As a result of these evasion strategies, relatively little interferon is produced during a rotavirus infection.

I: You're safe from the interferon system then?

RV: Not completely safe. It's very dangerous, that interferon, and no defensive strategy is fool-proof. Consequently, I have evolved another layer of protection. In addition to these schemes designed to avoid detection, I encode a protein, NSP1, which targets transcription factors IRF3 and IRF7 for degradation, blocking transcription of the genes for both IFN-α and IFN-β. My motto is: Leave nothing to chance!

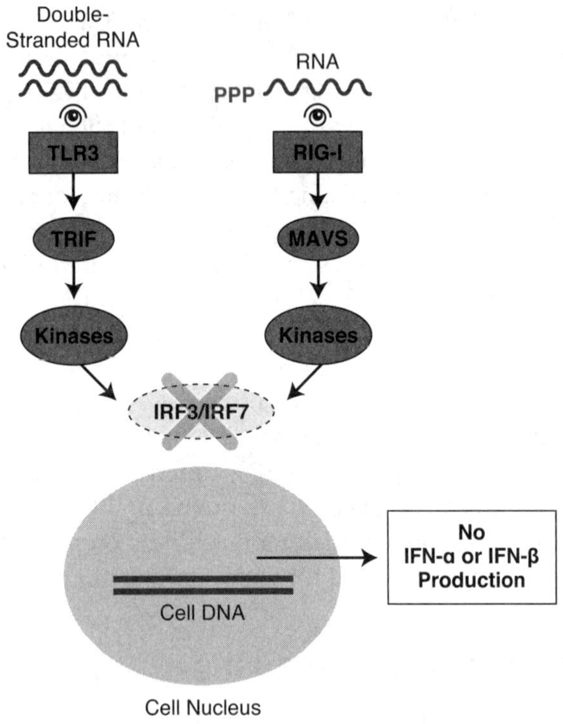

I: And the adaptive immune system? Any problems there?

RV: I've got that covered too. To avoid being destroyed by the powerful weapons of the adaptive immune system, I arrange to get my business done quickly and then exit. I'm a "hit-and-run" virus! The cells I infect can produce a large burst of new viruses in only about six hours. In contrast, it takes a week or more for the adaptive immune system to get going, and by that time many of the new rotaviruses will have exited with the feces to visit someone else. As a result, the main role for the adaptive system is to "mop up" any residual rotaviruses that the innate system (macrophages, complement, and a touch of interferon) haven't taken care of.

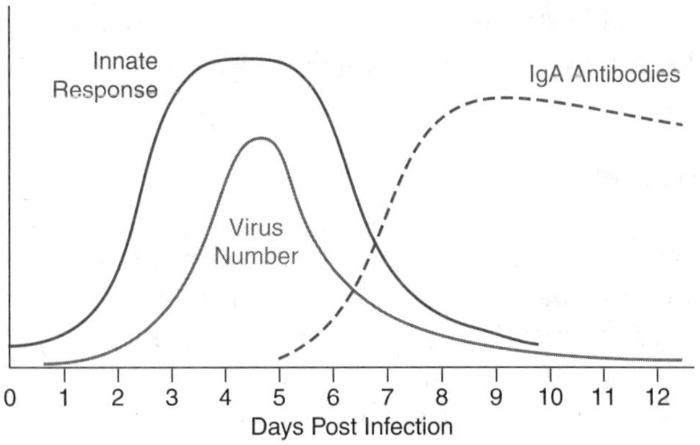

HOW ROTAVIRUS SPREADS

I: You mentioned feces, so I assume you use feces as a vehicle to spread to your new hosts.

RV: "Vehicle!" I like the way you said that. Very delicately put. Although relatively few intestinal cells are attacked during a typical rotavirus infection, these infected cells crank out so much virus that the stool (another delicate word) of an infected person can contain as many as one billion viruses per milliliter. And to be sure that these viruses "make it out safely," I encode a protein, NSP4, which acts as a toxin to induce diarrhea. Fortunately, my **virions** remain infectious while suspended in water, so I can be spread efficiently in a contaminated water supply. And because of my three capsids, I easily resist the effects of dehydration on dry surfaces. Consequently, I'm really good at just hanging out and waiting for the next human to pass by. As few as 10 rotavirus particles can initiate an infection, so, indelicate though it may be, I'm able to spread very efficiently by the fecal-oral route. Rotavirus is especially contagious among young children. Indeed, it is the rare four-year-old who has not been "visited" by a member of my family. After all, young children seem to produce feces almost continuously, and they like to put their mouths on everything in sight. Where would we be without those dear children?

I: Good point. So I understand that by speed and stealth, you manage to avoid the adaptive immune system long enough to reproduce and exit. But what about a return visit, as you call it? Can you infect the same human more than once?

RV: You can count on it! Because I reproduce so quickly, the immune system really only gets a glimpse of me as I "pass through," so a rotavirus infection frequently does not produce long-lasting immunity to a subsequent infection. In addition, my error-prone polymerase generates many mutations during replication, and some of these mutations can change the parts of my viral coat which are recognized by neutralizing antibodies. As a result of this antigenic drift, there are always several different rotavirus serotypes circulating in the population. Although rotaviruses can infect many different animals, including dogs, cats, and horses, none of these animals can efficiently pass type A rotavirus to humans. So I really don't have an animal reservoir. Nevertheless, like influenza virus, my genome is segmented, and humans can pick up and incorporate segments of animal rotaviruses. Consequently, I can use both antigenic drift and shift to "diversify" my genome and evade the adaptive immune system. Happily, because of speed and diversification, older children and adults frequently can be reinfected.

THE PATHOLOGICAL CONSEQUENCES OF A ROTAVIRUS INFECTION

I: Rotavirus infections frequently result in diarrhea, vomiting, and fever. I can understand why diarrhea might be helpful in spreading your infection, but why the vomiting and fever?

RV: Causing diarrhea is not just "helpful," it's essential. My whole deal is to reproduce under cover and get out quickly—and the diarrhea gets me out with a whoosh. My good friend, the hepatitis A virus, also uses the fecal-oral route, but he doesn't cause diarrhea. The reason is that he is non-cytolytic and reproduces in a rather leisurely infection. So it's okay for him if the newly made viruses just "drizzle" out. He's got plenty of time.

I: Okay, but what about the vomiting and fever?

RV: I do cause diarrhea intentionally to facilitate my spread, but the vomiting and fever is a different story. Vomiting involves a "reflex loop" which begins when nerves that have their inputs in the walls of the gastrointestinal tract sense that the gut is inflamed as a result of host defenses trying to kill me. And the fever is caused by inflammatory cytokines such as interleukin-1, which travel from the site of infection to the brain, where they trigger an increase in body temperature. I'd prefer not to make those kids vomit, and I don't like the fever any more than you do. But I didn't design your immune system. That's your deal.

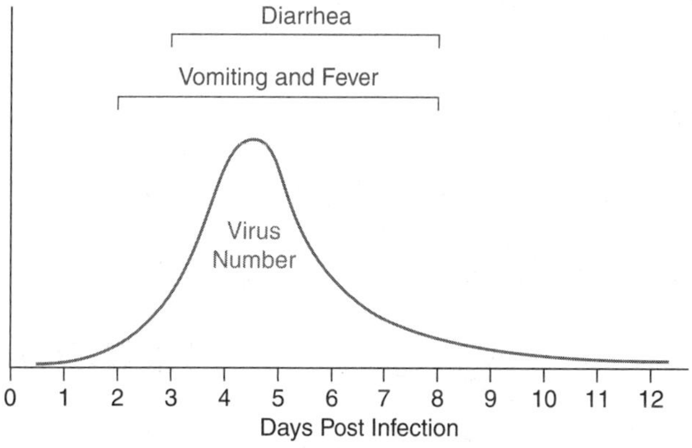

THE INTERVIEWER'S SUMMARY

Rotavirus cloaks its segmented, double-stranded RNA genome in three protein shells to protect it from the harsh conditions of the digestive tract. As a result, this virus is so "at home" in this environment that it commandeers the function of an intestinal enzyme to prepare it for entry into its target cells—the villous epithelial cells which line the intestine. The virus enters these cells via receptor-mediated endocytosis.

Rotavirus virus encodes a protein that interferes with the translation of polyadenylated cellular mRNA. By inhibiting the synthesis of cellular proteins, the virus replicates quickly, and produces a large number of new viruses per infected cell. Because its double-stranded RNA genome would immediately activate the interferon system, this clever virus copies its genetic information into single-stranded RNA while it is still inside two of its shells. Later, single-stranded viral RNA is enclosed in an inner protein shell, and only then does copying of single-stranded RNA segments take place to produce a new, double-stranded genome. This strategy of replicating "under cover" helps the virus avoid being detected by cellular sensors that are on the lookout for double-stranded RNA. Then, to be doubly safe from the interferon defense, this wily virus produces a protein that degrades transcription factors required for interferon production.

Rotavirus reproduces rapidly, kills the cells it infects, and only causes an acute infection. Because of its hit-and-run lifestyle, the adaptive immune system usually is sub-optimally activated during a rotavirus infection. In addition, the virus' error-prone polymerase can produce variant strains by antigenic drift, and its segmented genome makes it possible for the virus to exchange gene segments with animal rotaviruses, resulting in antigenic shift. Consequently, the immunological memory of the host's encounter with this virus is not long-lasting, and rotaviruses can "return" to reinfect older children and adults.

A viral protein induces diarrhea, which ensures that newly made viruses are quickly "washed out" of the host before they can be destroyed by host defenses. Rotavirus remains infectious in water, and the structure of its triple capsid makes it resistant to desiccation, so it has the perfect characteristics to make it a successful enteric virus.

GENERAL PRINCIPLES

1. Double-stranded RNA viruses and negative-strand RNA viruses must carry their polymerase molecules with them into the cells they infect. Positive-strand RNA viruses produce their polymerases after they have infected their target cells.
2. Enteric viruses must somehow protect themselves from the acid pH of the stomach and the digestive enzymes present in the small intestine.
3. Enveloped viruses usually are sensitive to the low pH found in the stomach. In addition, the bile in the intestine acts as a detergent, which can destroy the envelopes of most enveloped viruses – because their envelopes contain lipids. Consequently, the alimentary tract is not the preferred route of entry for enveloped viruses.

THOUGHT QUESTIONS

1. Why doesn't the rotavirus use its plus strand for protein synthesis? Why must it first copy its negative strand to produce mRNA?
2. How does this virus evade the interferon system?
3. How does rotavirus deal with the adaptive immune system?
4. Why do so many viruses disrupt cellular protein synthesis?

Adenovirus: A Virus With a Time Schedule

BACKGROUND

The second enteric virus we will discuss is the human adenovirus. In addition to human adenoviruses, there are adenoviruses that infect many kinds of birds and animals. Even frogs get adenovirus infections. So far, however, there have been no reports of adenovirus transmission between humans and other species (including frogs!). Human adenoviruses comprise a large family of viruses with about 50 different serotypes. In this chapter we will concern ourselves with two serotypes, 40 and 41, which cause gastrointestinal disease—the enteric adenoviruses. These adenoviruses are second only to rotavirus as the most frequent cause of infantile diarrhea, and by the age of three years, most children have been infected by an enteric adenovirus.

Adenovirus actually got its name because it was first isolated from human adenoid tissue, and many adenovirus serotypes do cause upper respiratory infections. Some adenovirus strains, for example serotypes 4 and 7, can infect both the respiratory tract and the gastrointestinal tract. Immunologists take advantage of the "dual targets" of these two serotypes when they prepare vaccines to protect army recruits from respiratory infections caused by adenovirus. Such infections could be a serious problem for troops living in close quarters. What is interesting is that the vaccine they use is made by packaging live adenovirus 4 and 7 in gelatin capsules, which are then swallowed by the recruits. Administered in this way, the viruses in the vaccine bypass the respiratory tract, where they would cause acute respiratory disease, and go on to establish an asymptomatic, immunizing infection of the epithelial cells of the small intestine. What is so elegant about this vaccination strategy is that the immunity generated by this asymptomatic intestinal infection protects not only against future intestinal infections, but also against respiratory infections by serotypes 4 and 7.

AN ENTERIC ADENOVIRUS INFECTION

Interviewer: I understand that you are an enteric virus just like the rotavirus I recently interviewed.

Adenovirus: No, I am not just like rotavirus. While it is true that rotaviruses and enteric adenoviruses are spread by the fecal-oral route, and infect the epithelial cells that line the intestine, we do things very differently.

I: Rotavirus has three protein coats. Since you also enter via the digestive tract, I presume you protect yourself with three coats too.

AV: No, I am not like rotavirus. I have only one protein coat. It's all I need to resist the low pH and detergents as I make my way through the digestive tract to reach my target cells in the intestine. That rotavirus is nothing but a show-off with those three coats. His extra coats are overkill. Actually my single capsid is quite handsome—rather like a soccer ball with "fiber" proteins sticking out. Some say I resemble a communications satellite.

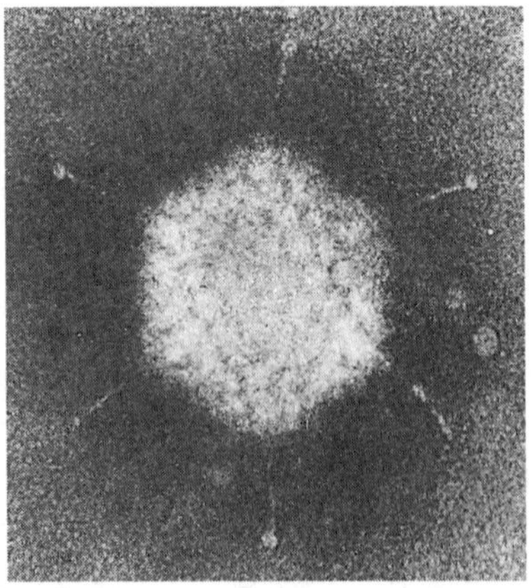

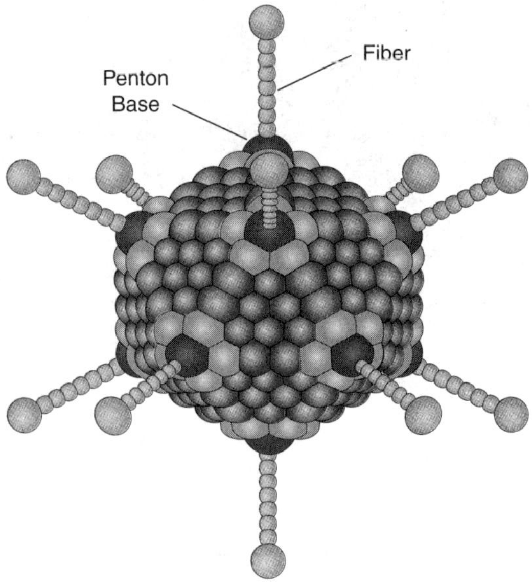

I: Yes, I can see what they mean.

AV: As you might predict, I use the knobs at the end of my fiber stalks to plug into fiber receptor molecules (CAR) on the surface of the epithelial cells that line the intestine.

I: Many human cells have those same receptors on their surface, yet enteric adenovirus infections do not spread beyond the digestive tract. Please tell us how you manage to limit your infection to the intestinal epithelium.

AV: Binding to the fiber receptor is not enough to permit me to enter a cell. Proteins on my surface also must plug into co-receptor molecules on the target cell surface. Most cells in the body do not have both the needed receptor and co-receptor molecules on their surface. It is only after the second contact with the co-receptor is made that I am taken into the cell, enveloped by an endosome.

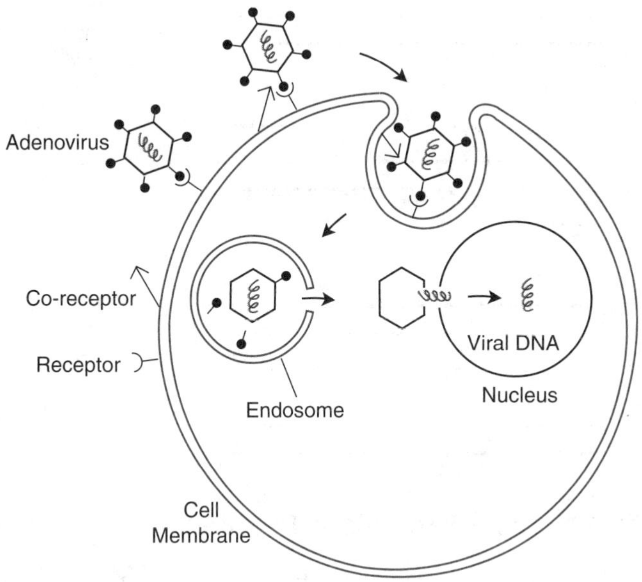

ADENOVIRAL REPRODUCTION

AV: Although I have a rather simple coat, my linear, double-stranded DNA genome has enough genetic information to encode more than 30 proteins—way more than rotavirus. These "extra genes" are what make me such an interesting virus. After my capsid has been partially dismantled by the acidic conditions within the endosome, I escape into the cytoplasm, and inject my cargo of DNA into the cell's nucleus.

I: That's odd. I would think that if your coat is "partially dismantled" in the acidified endosome, you would be stripped totally naked by the acidic conditions in the stomach. How do you accomplish this amazing feat?

AV: Sorry. That's for me to know, and some clever virologist to find out. However, I can tell you that once my DNA reaches the nucleus, cellular enzymes begin to transcribe some of my genes into mRNA. I say "some of my genes" because transcription of adenoviral genes proceeds according to a carefully orchestrated program—with certain viral genes being transcribed early after infection, and others being transcribed at later times. I am quite proud of this, because it is my ability to function on a strict time schedule which makes me such an effective virus.

I: Each time a cell divides to make two daughter cells, the DNA of the cell must be copied so that each daughter will receive a complete set of genetic information. Because these cellular "copy machines" are already in place, you DNA viruses can use the cellular DNA replication machinery to help copy your genome. Because both the adenoviral genome and the cell's DNA are linear and double-stranded, I suppose you just use the cell's DNA replication machinery "as is." After all, why reinvent the wheel?

AV: Why reinvent the wheel, you ask? I'll tell you why. No virus exists which uses exactly the same strategy for DNA replication as human cells do—it just won't work for viruses. One major problem has to do with the timing of DNA replication. When cells proliferate, cellular DNA replication is carefully controlled so that each origin of replication is used only once during a cell division cycle. This ensures that each chromosome is copied only once, and that each of the two daughter cells receives one complete copy of the genetic code—and no more. This scheme works just fine for cells, but if a virus used this strategy, only one new virus could be produced per cell cycle. We viruses make our living by using the cells we infect to make thousands of copies of our genome. So one copy per cell division just won't cut it.

REPLICATION OF ADENOVIRAL DNA

I: So DNA viruses like adenovirus must somehow uncouple their replication cycle from that of the cells they infect. How do you accomplish this?

AV: The way I do it is actually quite clever. To solve this problem, I encode my own DNA polymerase. This allows me to replicate in a way which is totally different from the way cells replicate their DNA. When cellular DNA is copied, a cellular DNA polymerase moves along one parental strand constructing a continuous, complementary daughter strand. At the same time, another DNA polymerase molecule copies the other parental strand. However, because the cell's DNA polymerase only works in one direction, copying this second parental strand results in small pieces of DNA, which must subsequently be joined together. So replication of cellular DNA is continuous on one strand and discontinuous on the other.

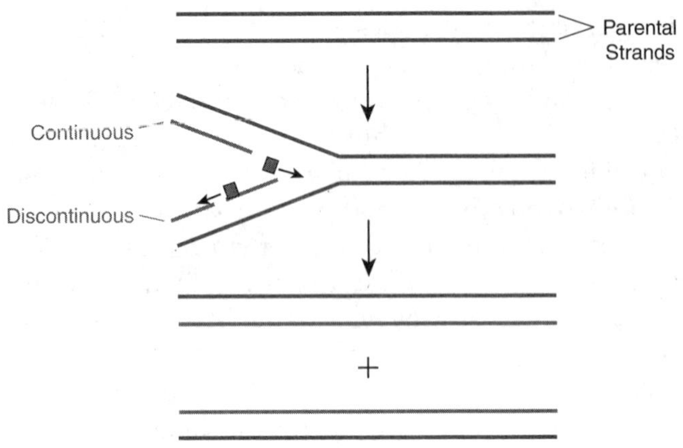

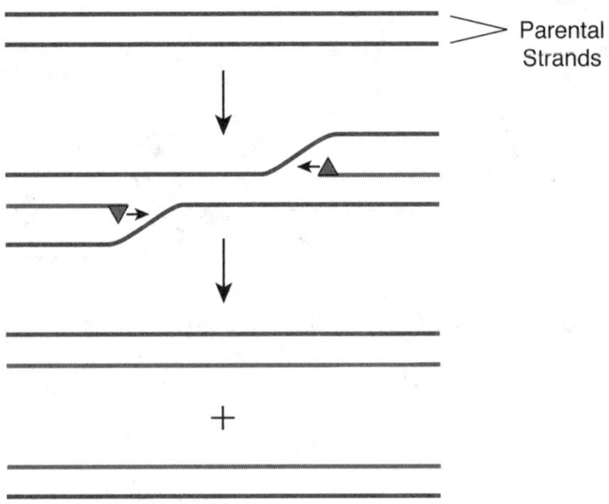

In contrast, when I replicate my DNA, my polymerase makes a complementary copy of one parental strand, displacing the second parental strand. The result is one double-stranded viral DNA molecule plus the displaced single strand. Another of my polymerase molecules then latches onto this displaced parental strand and copies it to make a second, double-stranded molecule. With this scheme, replication of both strands of viral DNA is continuous.

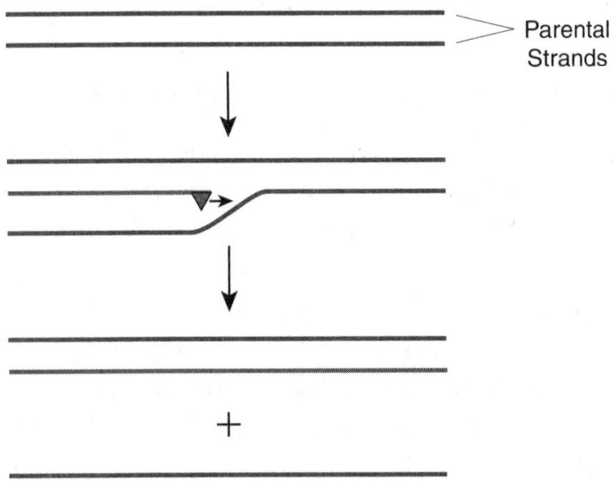

Another difference between cellular and adenoviral DNA replication is that the cellular DNA polymerase requires a short piece of RNA to "prime" DNA synthesis. In contrast, my replication is primed by viral proteins which bind to the 5′ ends of the double-stranded DNA. And here's the clever part. Because adenoviral DNA is symmetrical as far as the protein primers are concerned, DNA replication can actually begin at either end. Consequently, two polymerase molecules can roar along my DNA in opposite directions, simultaneously copying both strands.

The "Lazy Target Cell" Problem

I: That does seem efficient. However, your replication scheme requires not only several viral proteins, but also many of the supplies that the cell normally would use for replication of its own DNA. These include proteins involved in the copying operation as well as the nucleotide building blocks that are hitched together to form new DNA molecules. Doesn't this dependence on cellular replication factors cause a problem for you? After all, most cells in a mature human (e.g., the cells that make up my heart) are no longer actively replicating their DNA and dividing. They are "resting." And when the cellular copy machines of **resting cells** are not in use, they usually are shut down to conserve energy. Moreover, when cells are in a resting state, they generally don't keep large quantities of the materials needed for constructing DNA molecules on hand. So just how do you get the supplies you need to support your rapid replication scheme?

AV: Ah, yes. We DNA viruses call this the "lazy target cell" problem. To solve it, we have to trick the cell. Immediately after infection, we adenoviruses produce a protein (E1A) that triggers cellular DNA synthesis in the infected cell. When this happens, the cell begins to stockpile the materials we will need for replicating our genomes. Then, after these supplies are on hand, we start synthesizing mRNAs which encode our DNA polymerase and the other proteins required for viral DNA replication. The delay in producing these mRNAs makes sense: There's no reason for our polymerase to begin replication until the materials needed for the job are available. That would just lead to frustration. In fact, about eight hours elapse between the time we first enter a cell and the moment when synthesis of new viral DNA begins within the cell nucleus.

The Takeover of Cellular Protein Synthesis

I: Okay, so you trick the cell into providing the raw materials you need to replicate your DNA genome, but what about protein synthesis? To construct each adenoviral capsid requires over 1,600 protein molecules, and each infected cell produces a huge number of your virions. That requires a lot of viral protein synthesis! So how do you compete with the cell for the use of its protein-making machinery?

AV: Good point. We adenoviruses have two ways of taking over cellular protein synthesis and directing it toward production of viral proteins. Soon after infection, we make two proteins (the E1B-55K and E4orf6) which interfere with the transport of cellular, but not viral, mRNAs out of the nucleus. This focuses protein synthesis on viral transcripts, and ensures that viral proteins with enzymatic functions (e.g., my viral polymerase) are available early on. Then, later in the viral infection, when many proteins are needed to build thousands of viral capsids, we use another tactic: We rig things so that translation of viral mRNAs is heavily favored over translation of cellular mRNAs. To do this, we synthesize a protein (L4-100K) which interferes with the rapid initiation of cellular mRNA translation. Late adenoviral mRNAs are not affected by this inhibition because they have a special tripartite leader sequence at their 5′ end which "shunts" the 40S ribosomal subunits farther down the mRNA to the initiating AUG. The result is that during the final hours of an adenovirus infection, over 90% of the protein synthesized is viral.

I: Pretty impressive.

AV: Yes, and did you notice the careful timing? All this trickery relies on the carefully ordered expression of viral proteins. Adherence to this "time schedule" makes adenovirus replication so efficient that a single infected cell can make as many as 100,000 virus particles. This is 10 to 100 times as many virus particles as are produced by cells infected with most other viruses.

I: So everything must be precisely timed. But how do you do that?

AV: Well, you see, each adenovirus has a tiny little Rolex... No, not really! It's all done with complicated genetic circuits and feedback mechanisms. But that would just be too much information.

HOW DOES ADENOVIRUS EVADE HOST DEFENSES?

I: The careful timing makes sense, but doesn't it take several days from when you enter a cell until most newly made viruses have exited? It would seem to me that this leisurely pace of reproduction would make adenovirus-infected cells vulnerable to attack by host defenses. So I'm betting that you take effective countermeasures to ensure that viral reproduction can be completed before infected cells are destroyed.

AV: Lots of countermeasures! It might seem, for example, that an adenovirus infection would not induce interferon production. After all, I'm a DNA virus, and I don't replicate through a double-stranded RNA intermediate. However, my mRNAs are transcribed from both of my DNA strands, and some of these coding regions overlap. I've arranged it that way so that the same stretch of DNA can produce two different mRNAs—one from one strand and another mRNA from the other, complementary strand. It's a clever way of increasing the coding capacity of my genome. However, one consequence of this ploy is that mRNAs transcribed from these overlap regions have sequences which can base pair to produce long, double-stranded RNA molecules—which can be detected by cellular sensors.

I: So how do you protect yourself against the interferon system?

AV: Again, part of it is timing. Adenovirus inhibits cellular protein synthesis during what virologists term the "late phase" of infection, and this late phase starts when I begin to replicate my DNA. Up until that time, I've just been preparing to "get down to business"—the business of reproducing. The result of this careful timing is that by the time I transcribe the overlap stretches of viral DNA which produce double-stranded RNA, I am also producing proteins which inhibit cellular protein synthesis—which, of course, includes the synthesis of the interferon proteins. So by interfering with host protein synthesis, we adenoviruses reduce the amount of interferon which is produced.

This strategy isn't foolproof, however, and I also must defend myself against the effects of any interferon that is synthesized. One way I do this is quite unusual. Normally, in interferon-alerted cells, double-stranded viral RNA binds to the sensor protein, PKR, and activates this kinase, shutting down all protein synthesis, and ending the viral infection. To protect myself against this danger, my genome encodes "decoy" RNA molecules called VA RNA. This decoy binds to PKR and renders the protein kinase inactive, allowing protein synthesis to continue.

I: Making decoys to fool the interferon system is quite unusual. I wonder why other viruses don't do that?

AV: I don't mean to be immodest, but I think it's just a matter of brain power. If you are smart enough to operate on a precise time schedule, it isn't too difficult to envision using decoys.

The "Not So Fast, Buster" Defense

I: What about the adaptive immune system? Because you work so slowly, there should be ample time for the adaptive system to be activated before you have completed enough rounds of viral reproduction to efficiently infect a new host.

AV: You raise an important issue. During a first infection, the killer T cell is the adaptive immune system weapon we viruses fear most. Those killers recognize fragments of our viral proteins displayed by class I MHC molecules on the surface of infected cells. And when they see these signs of infection, they kill those cells (and the viruses within them!). It's pretty horrible. And because adenovirus makes so many different proteins in such abundance, you might expect that adenovirus-infected cells should be prime targets for destruction by killer T cells. Fortunately, we adenoviruses have evolved a mechanism that keeps viral proteins from being displayed on the surface of virus-infected cells.

Normally, class I MHC molecules are loaded with protein fragments in the endoplasmic reticulum from whence they proceed to the cell surface to display their cargo. However, my genome encodes a protein (E3-19K) which is anchored firmly in the endoplasmic reticulum. This protein grabs class I MHC molecules, and prevents them from traveling to the cell surface to display viral proteins. I call it my "not so fast, Buster" strategy. It works nicely, because if killer T cells can't see viral proteins displayed by class I MHC molecules on the cell surface, they have no way of knowing that the cell has been infected. Of course, the E3-19K proteins can't snag every single class I MHC molecule as it passes by, so this evasion strategy isn't perfect. But it does greatly decrease the destruction of adenovirus-infected cells by killer T cells, and it buys time for us to reproduce and spread before the adaptive immune system rids the host of all remaining viruses. After all, adenoviruses only cause acute infections which last a short time. We try not to overstay our welcome like some of those "freeloader" viruses that establish chronic infections.

Evading Apoptosis

I: All cells have "alarm systems" which alert them when biosynthesis is not going as planned. These alarms can initiate a cascade of events within the cell that leads to the cell's death by a process called "**apoptosis**"—a type of suicide. This sort of quality control is necessary because a cell's biochemical systems are so complicated that dysregulated cells are a frequent occurrence—and these out-of-control cells can pose a real threat to the human organism (e.g., by leading to cancer). You adenoviruses usurp the biosynthetic machinery of the cells you infect, so how do you prevent apoptosis from destroying you and the cells you infect before you complete your reproductive program?

AV: It's a cowardly act, this apoptosis. However, it is what it is, and we adenoviruses have to be ready to deal with it. Adenoviruses encode two, anti-apoptotic proteins, both of which are expressed early after infection. One protein (E1B-55K) blocks transcription of cellular genes that normally would activate the death program. The other (E1B-19K) binds to and inactivates key host proteins involved in initiating apoptosis. These apoptosis inhibitors make it possible for us to disrupt normal cell activities without the cell responding to the "something is terribly wrong" alarm.

The Death Protein

I: I can understand that suppressing apoptosis during an adenovirus infection is important—because a dead cell isn't going to produce much virus. But doesn't this suppression pose a potential problem? Although an adenovirus-infected cell is pretty beat up by the time viral reproduction is complete, unless something is done to really rip the cell open, won't the newly made viruses just slowly dribble out? Viruses like rotavirus, which reproduce quickly, actually use apoptosis to help accomplish their "final exit" from the cell. However, because you suppress apoptosis, you must have made some other exit arrangements.

AV: Enough already about rotaviruses! I'm sick of being compared to a trashy little piece of RNA that couldn't keep a time schedule if he had one!

I: Sorry.

AV: It's okay. I'm just a little sensitive. I get compared to rotaviruses all the time. But to answer your question: Yes, the final exit certainly would be a problem. However, late in infection, just when virus assembly is nearly finished, I synthesize a protein which virologists have named "the adenovirus death protein" (E3-11.6K). This protein acts to burst open the infected cell and allow the 100,000 new viruses trapped inside to come roaring out. So the E1B proteins, which are made early in infection, hold off cell death long enough for viral reproduction to be completed. Then, just when newly made viruses are ready to exit, the adenovirus death protein delivers the *coup de gras.* It sounds simple, but without strict adherence to a schedule of precisely timed gene transcription, it would be deadly. Just let your precious little rotaviruses try that one!

HOW DO ENTERIC ADENOVIRUSES SPREAD?

I: Enteric adenoviruses spread by the fecal-oral route, and young children are your main targets for infection.

AV: Yes, we do love children. Importantly, enteric adenoviruses are very resistant to drying, so that after the diarrhea the infection causes has spread the virus to the outside world, it can "lie in wait," fully infectious for long periods of time.

I: Nowadays there are plenty of kids around to infect, but do you ever infect the same human more than once?

AV: Yes I do.

I: Really? You interfere with the trafficking of class I MHC molecules, allowing you to hide from killer T cells. However, during a subsequent infection of the same host, how do you deal with the virus-specific antibodies which are made during the initial infection? It would seem that those antibodies would neutralize any invading adenoviruses before they could initiate a second infection.

AV: The answer is simple: I use antigenic drift. Many RNA viruses evade the antibody defense by mutating so that they can cannot be recognized by "outdated" antibodies produced during an initial infection. They can do this because their RNA polymerases are extremely error-prone, with no proofreading to correct these errors. My DNA polymerase is not as error-prone as a typical RNA polymerase, because it does have some capacity to proofread its work. Nevertheless, my DNA is not copied nearly as faithfully as is cellular DNA—which uses a polymerase that only makes about one mistake per billion bases. This is another reason why adenoviruses don't take the easy way out and just use the cellular DNA polymerase: By using my own polymerase to replicate my genome, I am able to generate antigenic drift. That's why there are so many human adenovirus serotypes.

PATHOGENIC CONSEQUENCES OF AN ENTERIC ADENOVIRUS INFECTION

I: I suppose that being an enteric virus, you cause the same symptoms as the rotavirus.

AV: There you go again! Rotavirus this, rotavirus that. Yes, we do cause many of the same symptoms—fever, vomiting, and diarrhea. Okay? But I don't just come in and blast away like that rotavirus. I take my time, because I am on a very sophisticated time schedule. In fact, about a week usually elapses between an adenovirus infection and the appearance of any symptoms. And then I "linger." After all, I have invested heavily in tactics that allow me to evade host defenses for a relatively long time. Why should I hurry?

I: Why hurry? Because I'm not a big fan of "eternal" diarrhea, that's why! At least the rotavirus gets it over with fast.

AV: May you be cursed with frequent rotavirus infections!

THE INTERVIEWER'S SUMMARY

The enteric adenoviruses spread by the fecal-oral route, infect and kill epithelial cells that line the intestine, and do not cause a disseminated infection. The adenovirus genome is a linear, double-stranded DNA molecule, enclosed in a single protein capsid. Transcription of viral genes takes place in a carefully ordered sequence. This "assembly schedule" ensures that adenoviral reproduction takes place efficiently, and that the different viral proteins are available exactly when they are needed.

Adenovirus enters cells by receptor-mediated endocytosis, and uses a novel DNA replication strategy to uncouple viral and host DNA replication. This makes it possible for many cycles of viral DNA replication to take place in the time it would normally take the cell to replicate its DNA only once. Because many of the cells the virus targets are resting, the virus encodes a protein that kicks these cells into "replication mode." As a result, the cell thinks it will be copying its own DNA and dividing, so it produces all the raw materials needed for viral reproduction.

Adenovirus makes two proteins which interfere with the transport of cellular mRNA out of the cell nucleus early in infection. This helps ensure that the early viral proteins are produced in a timely fashion. Then, later in the infection, when huge amounts of viral proteins are required to construct viral capsids, adenovirus encodes a protein which interferes with the translation of cellular messenger RNA, focusing protein synthesis on viral mRNA.

Adenovirus reproduces relatively slowly: About two days are required for an infected cell to produce a full burst of new viruses. As a result, it takes more than a week for this virus to infect enough cells to ensure that it can spread successfully to the next host. Because of this leisurely time schedule, the virus must protect itself against not only the host's innate defenses (e.g., interferon), but also against the adaptive immune system. To counter these defenses, the adenovirus produces a large number of "defensive proteins." In fact, roughly one-quarter of the adenovirus genome is dedicated to protective countermeasures. For example, adenovirus makes VA RNA which acts as a decoy to block the action of interferon-induced, antiviral proteins. Adenovirus also encodes a protein that interferes with the display of viral proteins by class I MHC molecules on the surface of infected cells, making these cells "invisible" to killer T cells.

An adenoviral infection profoundly disturbs cellular processes, and this dysregulation normally would trigger the cell to commit suicide by apoptosis. To prevent this, adenovirus encodes two proteins which disrupt the apoptosis program. Then, after many new viruses have been synthesized, adenovirus produces a "death protein" which facilitates the exit of these viruses from the infected cell—to be "set free" with the feces.

Eventually, the host's immune system deals harshly with an enteric adenovirus infection, and the virus is banished (cleared) from the host. And because the adenovirus doesn't "go gentle into that good night," the adaptive immune system becomes fully activated, and immunity to the infecting strain of adenovirus is long-lasting. However, the adenovirus DNA polymerase is error-prone—although not nearly as error-prone as most RNA polymerases—and this gives the virus the opportunity to mutate to produce different strains of virus, making reinfection possible.

GENERAL PRINCIPLES

1. The genomes of all DNA viruses are replicated in the infected cell's nucleus, where cellular DNA synthesis takes place.
2. No virus exists which uses exactly the same strategy for DNA replication as do human cells. DNA viruses must somehow uncouple their replication cycle from that of the cells they infect, so that multiple rounds of viral DNA replication can take place in the time it takes a cell to replicate its DNA once.
3. Viruses which do not reproduce quickly must suppress the cell's apoptosis (suicide) response at least long enough for them to reproduce and spread.
4. In an initial infection, the most important weapon the adaptive immune system can deploy against an attacking virus is the killer T cell. These killers can recognize viral proteins displayed by class I MHC molecules on the surface of infected cells, and can destroy those cells and the viruses inside them.
5. When a virus wishes to make a return visit to a person who was infected previously, the immune system's most powerful weapons are virus-specific antibodies – which can neutralize the virus before it can enter its target cells and initiate an infection.

THOUGHT QUESTIONS

1. Adenoviral genes are transcribed in a special temporal order. Give examples of why the orderly expression of its genes is essential to the adenovirus' lifestyle.
2. Adenovirus replicates its DNA in a way which is different from the way cells replicate their DNA. Describe this novel replication strategy.
3. There are at least two reasons why the adenovirus' novel DNA replication strategy is important for its lifestyle. What are they?
4. How does the adenovirus overcome the cell's apoptosis suicide defense?
5. How does the adenovirus defend itself against the interferon system?
6. How does the adenovirus defend against the adaptive immune system?

Chapter 9

Hepatitis A: A Virus That Detours

BACKGROUND

According to most estimates, over half the population of the United States has been infected with hepatitis A virus at one time or another, so we certainly need to include this one in our Parade. But what really makes this virus so interesting is that, although it enters its host through the mouth and exits through the anus just like rotavirus and the enteric adenoviruses, on its trip from top to bottom it takes a "detour" through the liver. It is this detour which makes hepatitis A virus such a successful human pathogen. It is also the main feature of the virus lifestyle that leads to the pathological consequences of a hepatitis A virus infection. Although several types of animals can be infected with hepatitis A experimentally, there is no known natural animal reservoir for this virus.

HOW DOES HEPATITIS A VIRUS INFECT ITS HOST?

Interviewer: You and the rhinovirus belong to the same family (*Picornaviridae*), and yet he is a respiratory virus, and you are an enteric virus. Why do you two "cousins" choose such different routes of infection?

Hepatitis A Virus: I've heard it said that "the clothes make the virus," and perhaps that's true. My protein capsid is resistant to the acid conditions in the stomach which would destroy the rhinovirus capsid. As a result of this difference in viral "clothing," rhinovirus must be content to be a rather ordinary respiratory virus, whereas I can be spread by the more glamorous fecal-oral route.

I: I'm not sure "glamorous" is the word I'd use to describe the fecal-oral route. However, because you are an enteric virus, I presume that your primary targets are the epithelial cells that line the intestine.

HAV: Yes, I do infect intestinal cells, but I infect very few of them, and these infections don't produce many new viruses.

I: That can't be the whole story. Otherwise you'd be in deep trouble in terms of persisting in the human population. So you must have some cunning tricks up your sleeve.

HAV: Indeed I do. My real target is the liver. That, of course, is why I'm called a hepatitis virus, *hepato* being Greek for liver.

I: I got that part, but how does a virus which enters the digestive tract manage to make its way to the liver and establish an infection there?

HAV: The story's a bit involved, but I think you'll enjoy it. The viruses produced from my initial infection of intestinal epithelial cells are taken up by "M" cells in the intestine. These specialized cells are part of the mucosal immune system—the arm of the immune system tasked with defending against invaders that enter the body via the digestive tract. The function of M cells is to sample the contents of the intestine, and to help initiate an immune response to potential invaders. Viruses and other invaders (e.g., pathogenic bacteria) bind to the surface of an M cell, and are transported through the M cell into the tissues below. From there the invaders are transported to nearby lymph nodes. In these nodes, B cells whose receptors recognize the virus become activated, and then proliferate to build up their numbers. This process takes a week or more. These selected B cells then travel back to the tissues underlying the intestine, and begin to pump out antibodies specific for the invader—in this case, me.

I: That doesn't seem clever at all, alerting the mucosal immune system.

HAV: Ah, but that's the genius of my approach! The antibodies made by these B cells are predominately of the IgA class—a class of antibody that is especially well suited to defend against intestinal invaders. IgA antibodies are "passive" antibodies, and their job is to bind to invading viruses and usher them out of the body—for example, by transporting the viruses back out into the intestine to be disposed of in the feces. Indeed, one of the constraints on the mucosal immune system is that it must deal with invaders without causing much inflammation. The reason is that your digestive tract is under continual attack by bacteria and viruses, and if the immune system caused inflammation in response to all these attacks, you'd have diarrhea all the time. And that wouldn't be good—at least not for you. So IgA is the antibody class of choice to defend against an intestinal infection.

In contrast, the other class of antibody which can defend against a viral infection, IgG, usually causes a lot of inflammation. Consequently, the B cells which protect the intestine generally do not make IgG antibodies. As you'll see, that's important.

I: That's all very interesting, but, again, why do you intentionally alert the immune system?

HAV: Don't be in such a hurry. Are you rushing off to interview some other virus? You need to have some background so you will fully appreciate how clever this strategy really is. Now, where was I? Oh, yes.

There is another way IgA antibodies usher viruses out of the body, and it is a bit more circuitous. IgA antibodies which are bound to viruses (antibody-virus "complexes") in the tissues that underlie the intestine are collected by the lymphatic system, poured into the blood stream, and sent to the liver. Liver cells have receptors for IgA antibodies, and once connected to these receptors, both the IgA antibodies and their cargo of invaders are taken inside the liver cell for disposal. Now the good part: It turns out that I actually use this IgA "disposal system" to solve my problem of how to infect liver cells! As IgA antibody-virus complexes collected from intestinal tissues travel to the liver for disposal, I happily go along for the ride (I'm the virus in the antibody-virus complex). Then, when we reach the liver, I use the uptake of these complexes to usher me right into the very liver cells I want to infect (the hepatocytes). Rather elegant, I must say.

I: So you actually "hijack" the immune system to take you where you want to go?

HAV: Exactly. I like liver cells because they have all the goodies I need in order to turn them into virus factories. Moreover, viruses produced in liver cells are released into the bile ducts which drain the liver, and then are emptied right into the small intestine with the bile. Of course, this all takes a while, and the "detour" through the liver and back to the intestine is the reason newly made viruses usually are not detected in feces until weeks after the initial infection. Actually, the return trip to the intestine is a piece of cake, because my protein coat is unfazed by the bile salts, which act as detergents—and which would destroy ordinary viruses. But then, I'm no ordinary virus!

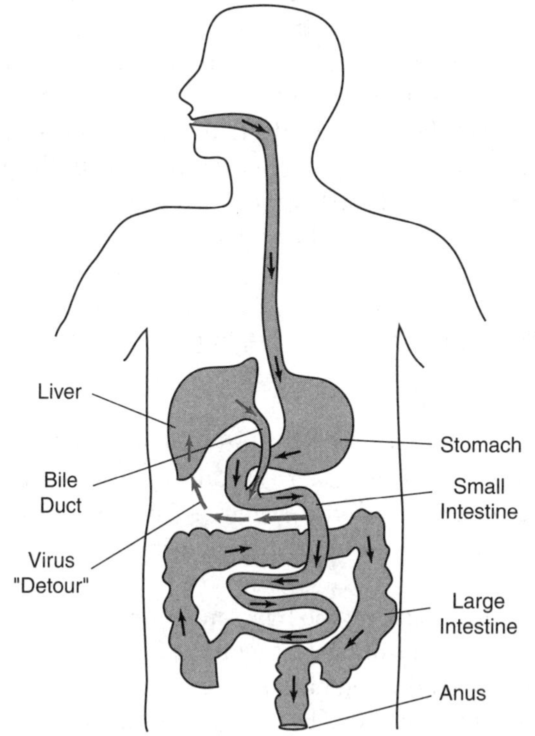

HOW DOES HEPATITIS A VIRUS REPRODUCE?

I: I can see that you are quite special, but you skipped over how you reproduce in liver cells. How does that go?

HAV: My genome consists of a single piece of positive-strand RNA encased in a single capsid made of protein—just like the rhinovirus. In fact, we reproduce in a similar fashion in the cells we infect.

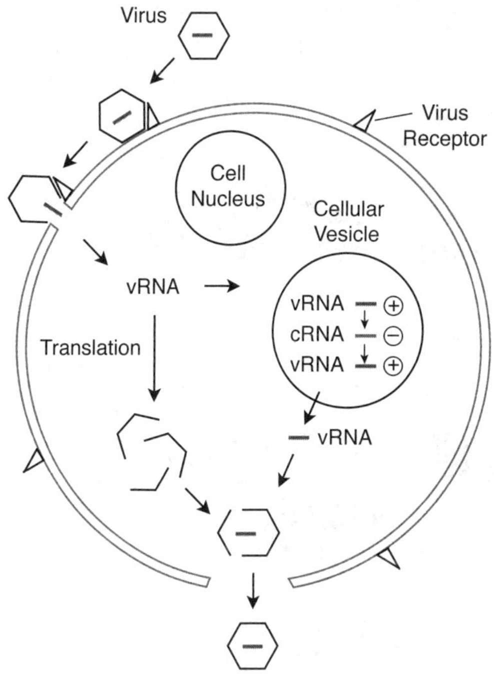

Although our reproductive strategies are similar, there are important differences in our lifestyles. For example, rhinovirus is a "hit-and-run" virus which shuts down synthesis of host proteins, and kills the cells it infects. Hepatitis A virus reproduces "slowly and gently," making new viruses without causing significant damage to infected cells. I'm much more of a gentleman than Cousin Rhino.

HOW DOES HEPATITIS A VIRUS EVADE HOST DEFENSES?

I: You mentioned being slow and gentle, so I'm sure you must have evolved ways of dealing with the interferon warning system.

HAV: Yes. In fact, I've never met a virus which didn't have to defend itself against that danger. Unfortunately, liver cells are littered with RIG-I pattern recognition receptors that can detect my uncapped viral RNA. And, of course, there are TLR3 molecules which can recognize the double-stranded RNA that is produced when I replicate. So I have my work cut out for me. To deal with the RIG-I problem, my genome encodes a protein, 3ABC, which destroys the important MAVS signaling protein that is required for RIG-I to activate the IRF3 transcription factor and turn on IFN-β production. In addition, I use one of my proteins (3CD) to cleave the adaptor protein, TRIF, so that it cannot pass along the signal from TLR3 and trigger interferon production. These ploys are so effective that very little interferon is made during a hepatitis A virus infection.

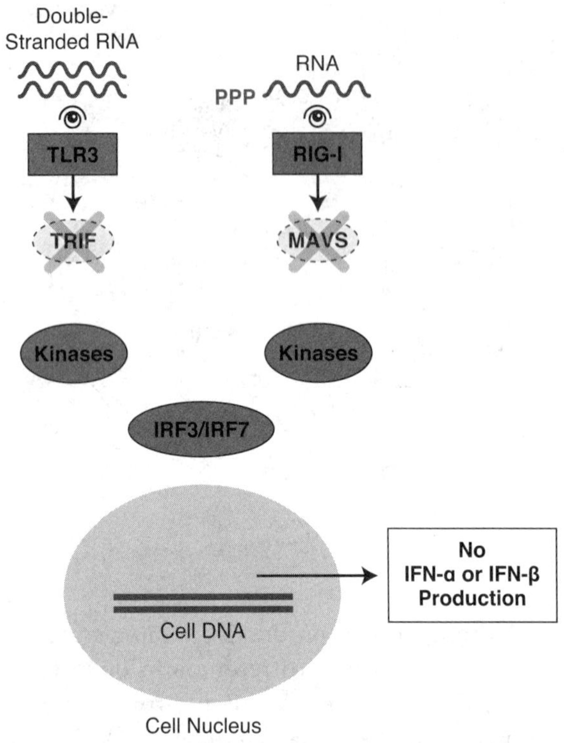

I: I would also expect that the adaptive immune system would pose a formidable problem for you because of the relatively long time which elapses between infection and virus production.

HAV: Not really. When the adaptive immune system which protects the intestine is alerted, its response is to produce virus-specific, IgA antibodies—the antibodies I use as a "vehicle" to facilitate my infection of liver cells. However, IgA antibodies are designed to protect the digestive tract, not the liver. The weapons which the immune system uses to deal with viruses that infect the liver are IgG antibodies and killer T cells. Consequently, once I reach the liver, the immune system must begin from scratch to produce those weapons. And during the week or two while the "liver immune system" is firing up, I'm reproducing like crazy in infected liver cells.

I: So, in a sense, the weapons of the adaptive system which are mobilized to deal with your initial infection of intestinal cells are "obsolete" once you detour to the liver.

HAV: My point exactly. Hepatitis A virus "confuses" the adaptive immune system by moving from the intestine, where IgA antibodies are effective, to the liver, where they are not. This evasion strategy works so well that it is three to five weeks post infection before the "liver arm" of the immune system is ready—and by then, I'm bathing in bile.

HOW DOES HEPATITIS A VIRUS SPREAD?

I: You noted that when you infect cells in the intestine, not much virus is produced. Is it the same way with liver cells?

HAV: Oh, no. I infect those intestinal cells gently on purpose—just enough infection to make a few more viruses and provoke the adaptive immune system into making the IgA antibodies I need to reach my real target cells. In contrast, a hepatitis A infection of liver cells is robust, producing large quantities of virus. In fact, at the height of a hepatitis A infection there can be 100 million virus particles excreted per milliliter of feces. That's a lot of virus!

I: I suppose that, eventually, the IgG antibodies and killer T cells generated in response to the liver infection are able to destroy all of the viruses that remain in the body.

HAV: Yes, that's true. The immune system mops up very well, so I never cause a chronic infection. However, by the time the immune system is fully engaged, most newly made viruses have already exited with the feces.

I: Because you target the liver, an organ through which blood flows continuously, I would expect that you also could be spread when humans exchange blood or blood products (e.g., during a blood transfusion or when drug abusers share needles).

HAV: I suppose that could happen. However, because I don't establish a chronic infection, the period of time when an infected individual's blood contains hepatitis A virus is generally so short that the probability of spread by blood-to-blood contact is quite small.

I: Can hepatitis A virus infect the same human more than once?

HAV: There is only one hepatitis A serotype, and infection generally leads to lifelong immunity. So I usually get to visit each host only once. Consequently, I depend on lax hygiene and a large population of susceptible individuals for my continued existence. But that's not a problem, because I am "built to last." Indeed, after four weeks in dried feces, my infectivity only decreases by about a factor of 100. In addition, I can survive for weeks in shellfish, which can concentrate hepatitis A virus by filtering large volumes of contaminated water. These features make me ideally suited to be spread efficiently by the fecal-oral route. Unfortunately, I am sensitive to the concentrations of chlorine commonly used for water treatment—and also to toilet bowl cleaner! But you'd be surprised how many people don't wash their hands after using the toilet.

VIRAL PATHOGENESIS

I: I suppose, like the other enteric viruses I interviewed, you cause diarrhea.

HAV: Would a gentlemanly virus do such a thing? Certainly not! After all, once I infect liver cells, there is no need for diarrhea. Newly minted viruses are emptied right back into the intestine, and simply join the normal "ebb and flow."

I: But don't people get sick at all when you infect them?

HAV: Some do, but most don't. In fact, young children rarely know they have been infected. Because the initial infection of cells in the small intestine is so gentle, the early phase of a hepatitis A infection is uniformly asymptomatic. A week or two into the infection, we arrive in the liver, and soon after that, viruses produced by infected liver cells begin to be excreted in the feces. After about another week, killer T cells, activated in response to the liver infection, start to appear. Their job is to destroy virus-infected liver cells—and they do their job very well. However, the liver is a big organ, and the number of hepatitis A–infected liver cells usually is too small to compromise liver function. I should point out that because I am not a cytolytic virus, the killing of liver cells results from the immune response to the infection, not from the viral infection itself.

In a minority of hepatitis A infections, mainly in adults, the destruction of liver cells is more extensive, and symptoms characteristic of viral hepatitis begin to appear about four weeks post infection. In the human body, about 100 billion aged red blood cells are "retired" each day. These effete cells are rapidly digested by macrophages, and the iron they contain is recycled. However, the part of the hemoglobin molecule which cradles the iron atom cannot be reused, and after it has been processed by the macrophage to form a yellow pigment called bilirubin, it is spit out into the blood or tissues surrounding the macrophage.

Because each red blood cell contains so many hemoglobin molecules, and because so many red blood cells are retired from service each day, the huge amount of bilirubin produced creates a major disposal problem. To deal with this, most of the bilirubin is complexed with proteins in the blood (primarily albumin) to make it soluble, and is carried to the liver. There the bilirubin is taken up by hepatocytes, modified, and released into the bile. When the system is working properly, disposal by the liver is so efficient that the concentration of bilirubin in fluids and tissues remains low. However, when liver cells are destroyed by the immune response to a hepatitis A infection, this disposal system can be overwhelmed. When that happens, bilirubin concentrations increase dramatically, resulting in jaundice and dark urine.

Although bilirubin is not terribly toxic, jaundice and dark urine are pretty good indicators that the liver is not functioning properly. Because the liver is tasked with detoxifying many other waste products of normal cellular metabolism, jaundice is usually accompanied by symptoms such as malaise, loss of appetite, fever, nausea, and vomiting—symptoms which are caused by inadequate detoxification of cellular waste products due to liver damage.

I: That doesn't seem all that "gentlemanly" to me.

HAV: I think it sounds worse than it really is. Hepatitis A virus never establishes a chronic infection, and the immune system—which, you remember, is causing all the damage anyway—usually requires only a few weeks to eradicate any viruses that get left behind. Moreover, liver cells destroyed by the immune response are quickly replaced by the proliferation of healthy liver cells. As a result, symptoms associated with a hepatitis A infection are generally short-lived, and I rarely cause life-threatening disease. In the United States, for example, there are only about 100 deaths each year associated with hepatitis A infections, and these are mainly in older age groups.

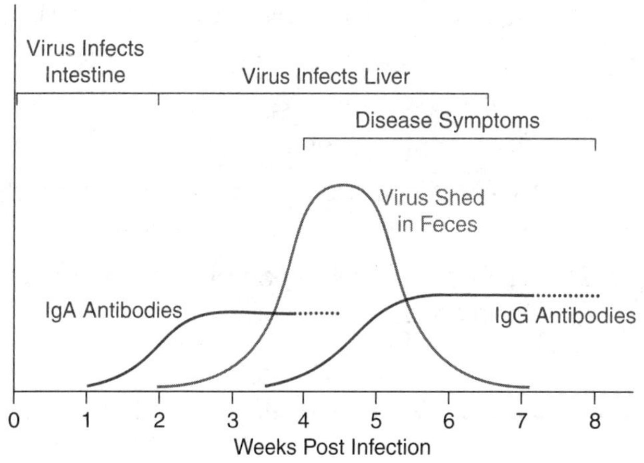

THE INTERVIEWER'S SUMMARY

Like rhinovirus, hepatitis A virus is a single-stranded, positive-sense RNA virus with a single protein capsid. Although rhinovirus and hepatitis A virus are members of the same family, and replicate their genomes in a similar fashion, rhinovirus is a respiratory virus, and hepatitis A virus is spread via the fecal-oral route. Hepatitis A virus' infection strategy is first to be "eaten," and then to initiate a limited infection in the intestine. This virus does not kill the cells it infects, and this "gentle" infection of intestinal cells tricks the immune system into making IgA antibodies. Hepatitis A virus then uses these IgA antibodies as a "taxi service" to take it where it really wants to go—to the liver.

Hepatitis A virus binds to its target cells, and its genome is released directly into the cytoplasm. It then uses an internal ribosome entry site to facilitate translation of its genome to produce a single, long polyprotein—which is subsequently cleaved to yield smaller viral proteins. When it infects hepatocytes, hepatitis A virus blocks the cell's ability to signal that it has been infected, so that very little interferon is produced. And because IgA antibodies are useless in the liver, the adaptive immune system must "reset" to make the weapons appropriate to deal with a liver infection—IgG antibodies and killer T cells. This trickery buys time for infected liver cells to produce copious new viruses, which are "naturally" carried with the bile back to the digestive tract to be excreted with the feces.

The bottom line is that hepatitis A virus takes a detour through the liver on its way through the digestive tract, and in doing so, evades host defenses long enough to become one of the world's leading pathogenic viruses. Once the adaptive immune system reaches full strength, the infection in the liver is easily dealt with, usually without significant damage to the liver. And because there is only one serotype of hepatitis A virus, humans cannot be reinfected.

GENERAL PRINCIPLES

1. Subtle changes in viral design can result in major differences in a virus' route of entry, and the diseases it causes.
2. Viruses that are spread by the fecal-oral route usually are resistant to drying, and can survive for long periods on objects.
3. Only viruses which establish a chronic infection are efficiently spread by exchange of blood or blood products.

THOUGHT QUESTIONS

1. In what important ways does hepatitis A's lifestyle differ from the other two enteric viruses we have discussed, rotavirus and adenovirus?
2. How does hepatitis virus make its way from the intestine to the liver – and back?
3. From the host's perspective, why is the liver a "good" organ for hepatitis A virus to target? And why does the virus pick this organ?
4. A hepatitis A infection does not cause diarrhea. Why is this advantageous for the virus (clue: Where do people with diarrhea usually hang out?)?
5. Only viruses which establish a chronic infection are efficiently spread by exchange of blood or blood products. Why?

Review 2

A Comparison of Enteric Viruses

In the last few chapters, we discussed three enteric viruses—heroic viruses which brave many perils to infect cells of the intestinal tract. First we "interviewed" the rotavirus, with its segmented, double-stranded RNA genome protected by three protein coats. This virus was contrasted with the human enteric adenovirus which has a large, non-segmented, double-stranded DNA genome and only one protein coat. Finally, we examined the life-style of hepatitis A virus—a single-stranded, positive-sense RNA virus with a single protein capsid.

All three viruses are spread by the fecal-oral route, infect epithelial cells that line the intestine, and cause only acute infections. Rotavirus and adenovirus use receptor-mediated endocytosis to enter their target cells. In contrast, once hepatitis A virus binds to cellular receptors, its genome is "injected" directly into the cell's cytoplasm. Although hepatitis A virus enters by the mouth and exits by the anus—just as rotavirus and the enteric adenovirus do, the infection of intestinal cells by hepatitis A virus is of relatively little consequence. Indeed, for hepatitis A virus to survive, it must infect liver cells—a feat it accomplishes in a rather devious manner. When hepatitis A virus infects cells in the small intestine, IgA antibodies are produced which can bind to the virus. These antibodies then travel through the lymph and blood to the liver, carrying their "cargo" of hepatitis A viruses. Consequently, this virus takes advantage of the immune system's response to the initial infection of intestinal cells to "hitch a ride" to the liver. Once in the liver, hepatitis A virus reproduces efficiently in hepatocytes, and the newly made viruses then leave the liver with the bile—which is poured back into the small intestine to exit with the feces.

Each virus employs a distinctive strategy for reproduction. Rotavirus carries out mRNA synthesis and genome replication "under cover," enclosed in at least one of its protein shells. The adenovirus employs a unique mode of DNA replication in which both DNA strands are replicated continuously. And hepatitis A virus uses an internal ribosome entry site on its mRNA to produce one long polyprotein. Whereas rotavirus infects cells and produces new viruses in only about six hours, adenovirus reproduces slowly, with over 30 different genes being expressed in a carefully controlled sequence. In fact, it takes several days for even the first adenovirus-infected cells to begin to produce virus. In contrast, a week post infection, the rotavirus has "left the building." A hepatitis A virus infection also is rather leisurely, in large part because this virus sequentially infects cells in the intestine and then in the liver. Rotavirus focuses the biosynthetic machinery of infected cells on viral synthesis by blocking the translation of polyadenylated cellular mRNA. Adenovirus achieves a similar result by first tricking the infected cell into preparing to proliferate, and then by interfering with host protein synthesis both early and late in infection.

To evade destruction by host defenses, these three viruses also use very different strategies. Although rotavirus is a cytolytic virus, it is essentially a "pacificist" which tries to avoid the interferon defense by replicating within its protective coats, and by encoding a protein that degrades transcription factors required for interferon production. In addition, rotavirus reproduces so quickly that it leaves the infected host before the adaptive immune system becomes fully activated—being cleared from the body mainly by the action of the innate immune system. Consequently, immunity to a rotavirus infection usually is not complete, and a person can be infected again later in life by the same rotavirus serotype. In contrast, adenovirus is an "activist." By cleverly timing the expression of its many genes, adenovirus completely takes over its host cell, foils the cell's apoptotic response to the takeover, thwarts the interferon system by blocking its effects, and resists destruction by killer T cells by interfering with the display of viral proteins by class I MHC molecules. Adenovirus also produces a "death protein" at exactly the right time to "explode" the infected cell, and facilitate the rapid exit of about 100,000 newly made viruses. Fortunately for its host, the adaptive immune system eventually does destroy this cytolytic virus, and immunity against the infecting adenovirus serotype is long-lasting.

Hepatitis A virus employs a different set of evasion techniques. This virus produces a protein that disrupts the signals sent by pattern recognition receptors which detect the viral infection. Consequently, very little interferon is produced during a hepatitis A infection. This virus also uses a unique trick to confuse the adaptive immune system. When hepatitis A virus infects cells in the intestine, the adaptive system responds by producing IgA antibodies—antibodies which are designed to be used against viruses that infect the intestine. However, hepatitis A virus then turns the immune system against itself by using these very same IgA antibodies to "chauffeur" it to the liver. And once the virus infects liver cells, the immune system must begin all over again to make IgG antibodies and killer T cells—weapons that are effective against viruses that infect the liver. This "rebooting" of the immune system takes time, and allows the virus to reproduce in the liver and exit the body before the immune system reaches full strength. Rotavirus and the enteric adenovirus also can confuse the adaptive immune system by using their error-prone polymerases to create antigenic drift, and the rotavirus' segmented genome makes antigenic shift possible for this virus.

Rotavirus and adenovirus both can cause fever, vomiting, and diarrhea. Thus, two viruses which have evolved to live very different lifestyles can spread by the same route, infect the same cells, and cause many of the same disease symptoms. However, because the rotavirus is a "hit-and-run" virus, the symptoms which accompany an enteric adenovirus infection appear later, and generally last longer than those of a rotavirus infection. The "detour," which hepatitis A virus takes on its way from mouth to anus eliminates the need for this virus to cause diarrhea, since new viruses produced in the liver are delivered directly to the intestine to be released "naturally." However, this detour is responsible for the disease symptoms sometimes associated with a hepatitis A infection. Although hepatitis A virus usually does not kill the cells it infects, destruction of infected liver cells by the immune response can compromise liver function. Fortunately, hepatitis A virus never establishes a chronic infection, and the virus usually is eliminated by the immune system before much liver damage can occur.

Chapter 10

VIRUSES WE GET FROM MOM

Hepatitis B: A Decoy Virus

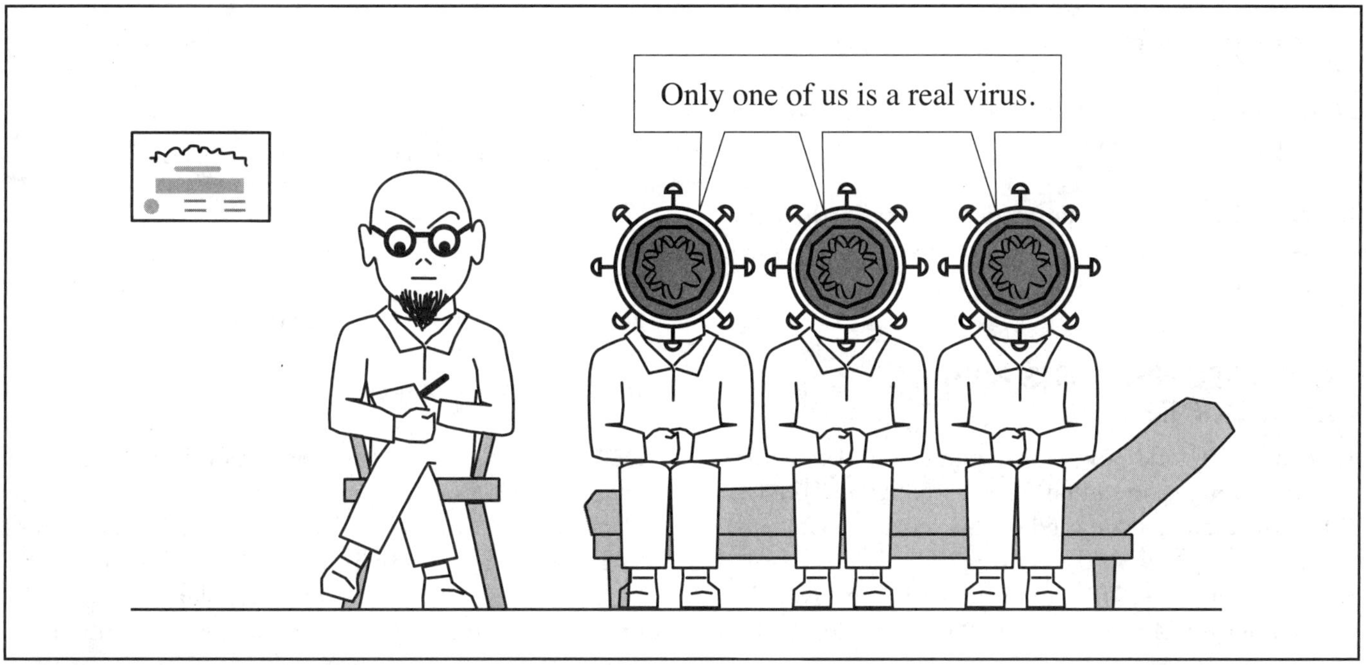

BACKGROUND

Hepatitis B, hepatitis C, and human T cell lymphotropic virus type I (HTLV-I) are three viruses which can be spread from mother to child. Although these viruses have very different lifestyles, they have one important common feature: All are able to establish lifelong, chronic infections. This makes perfect sense for a virus which spreads from a mother to her baby, because the virus must persist in the body of the infected child until she, too, can become a mom.

Hepatitis B virus has the smallest genome of any human virus, with only about 3,200 base pairs of genetic information. Compared to adenovirus, which has a DNA genome of roughly 35,000 base pairs, hepatitis B virus is a genetic runt. Yet despite its limited coding capacity, hepatitis B virus is one of the world's most successful human pathogens, with about 400 million people carrying this virus as a chronic infection.

HOW DOES HEPATITIS B VIRUS INFECT ITS TARGET CELLS?

Interviewer: Since you are a hepatitis virus, I presume that you target cells of the liver, and that you reach these cells via the blood. However, the bloodstream is a rather inhospitable place for most viruses, so you must be specially adapted to use this route of infection.

Hepatitis B Virus: Yes, you gotta be tough to survive in the blood. Fortunately, I have a lipoprotein envelope which is perfectly suited to resist the assaults of enzymes that are found in the blood. As a result, large quantities of hepatitis B viruses can accumulate over time in the blood of an infected individual.

I: You mentioned your envelope. Is that the only "wrapper" you have for your genome?

HBV: No. My genetic information is protected by both a protein capsid and a surrounding lipid envelope. When proteins on my envelope bind to receptors on the surface of liver cells, my envelope fuses with the cell membrane, and the encapsidated genome is released into the cell's cytoplasm. There my viral capsid is removed, and my genome enters the cell's nucleus.

I: You didn't mention which cellular receptor proteins you use for attachment and entry.

HBV: No, I didn't. That information is "classified." Virologists have been trying for years to figure it out. They can keep on trying.

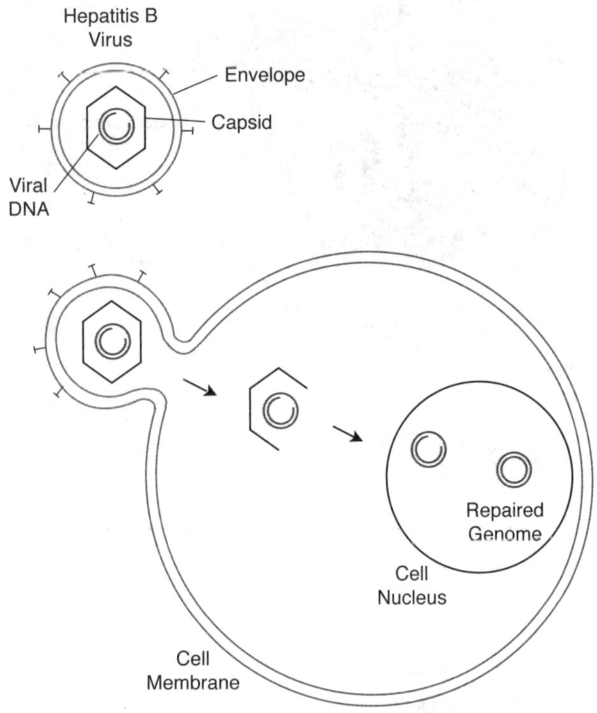

HOW DOES HEPATITIS B VIRUS REPRODUCE?

I: I can understand how information about the cellular receptors you use might be "used against you," but can you at least tell us what kind of genome you have?

HBV: Yes, of course. That information is even available on the Internet. I have a circular, DNA genome which is mostly double-stranded.

I: Did you say a "mostly" double-stranded DNA genome?

HBV: Yes, the hepatitis B genome has a gap where the DNA circle has only one strand. It's one of my most famous features. When my genome reaches the nucleus, a polymerase "repairs" this gap to produce a completely double-stranded, circular DNA molecule.

HBV: I'll bet you want to ask me how I end up with a genome that is mostly double-stranded.

I: Yes, that does sound rather strange.

HBV: I like to think of it as unique, not strange. Anyway, once my genome has been "repaired," so that it is completely double-stranded, I use the cell's RNA polymerase to transcribe the negative strand into mRNAs of various lengths. These mRNAs are then translated to make the proteins required for viral reproduction. In addition, the cell's RNA polymerase also makes full-length, complementary copies of my negative DNA strand to produce "genomic" RNAs.

I: So far, that doesn't sound too strange.

HBV: Unique. But here's where it starts to get interesting. After this genomic RNA has been transported out into the cytoplasm, the proteins that will form the viral capsid begin to assemble around it.

I: Wait! You are a DNA virus, not an RNA virus with a single-stranded genome. Why would you encapsidate genomic RNA? It's as if you can't make up your mind whether you want to have an RNA genome or a DNA genome. Are you confused?

HBV: Confused? Certainly not! As the capsid is being constructed, my personal reverse transcriptase enzyme (the cell doesn't have one) begins to make a complementary DNA copy of the genomic RNA, degrading this RNA after it has been transcribed. Then, once this process is complete, the reverse-transcribed DNA strand is used as a template by my reverse transcriptase to make a viral genome composed of double-stranded DNA. Now to answer your question: During synthesis of the second DNA strand, my polymerase must race to finish the job before the viral capsid is completed around it—and capsid assembly always wins! The result is a DNA genome which is only partly double-stranded.

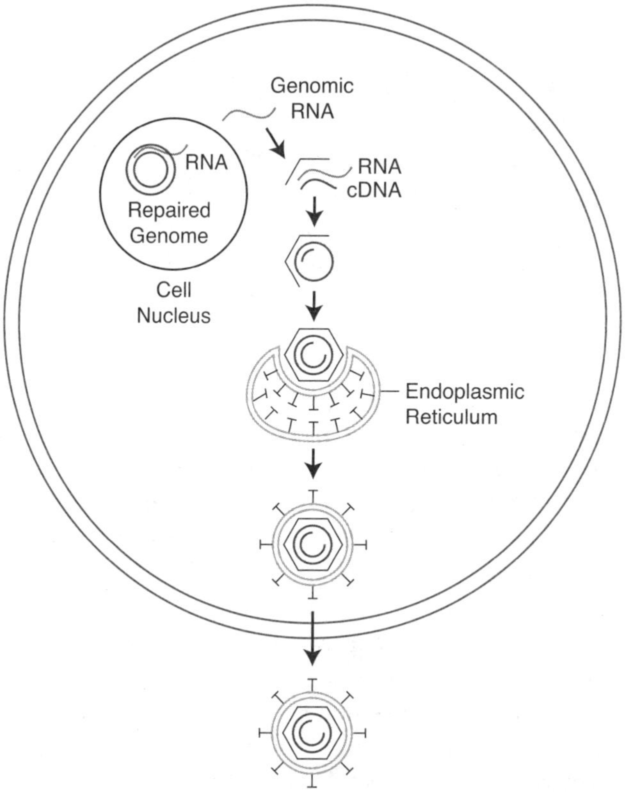

I: I would say that you easily win the award for "Virus With the Most Bizarre Replication Strategy"!

HBV: Please. The word is unique! No other virus replicates this way, so you can be sure that hepatitis B virus arose from a unique evolutionary event. Plagiarism is not our thing.

I: No, I guess not! So when viral assembly is complete, what happens next? Any more "unique" features?

HBV: Yes indeed. My genome encodes several proteins (e.g., the major surface antigen, HBsAg) which are inserted into the membrane of the endoplasmic reticulum of the infected cell. There these proteins assemble to produce viral envelopes that bud into the interior of the endoplasmic reticulum. If a protein capsid containing viral DNA binds to the outside of the endoplasmic reticulum when one of these envelopes is assembling, it will be included inside the envelope, transported to the surface of the cell, and released into the surrounding tissues. However, it really isn't important that every envelope contains a completed, DNA-containing capsid. In fact, a hepatitis B–infected liver cell produces about a thousand times as many empty viral envelopes as full.

I: My, isn't that wasteful?

HBV: No, not at all. Those empty virus particles play a major role in my ability to evade host defenses.

HOW DOES HEPATITIS B VIRUS EVADE HOST DEFENSES?

I: That brings me to my next question. What are your evasion tactics?

HBV: I have quite a few, so I'll just mention a couple of the most important ones. You kidded me about my unique replication strategy, but replicating within my capsid helps avoid detection by cellular sensors that would otherwise trigger the interferon defense. In addition, two of my viral proteins, the polymerase protein and the X protein, function to block the signals from cellular pattern recognition sensors which normally would trigger interferon production.

Another "deceptive" feature of my reproductive style is that although infected liver cells can produce large quantities of virus, these cells usually are not killed by the viral infection. In fact, I infect my target cells so gently that when a person is first infected, almost two months can elapse before a significant amount of virus is produced. Non-cytolytic viruses like me present a major problem for immune surveillance. This difficulty arises because, before the adaptive system can be activated to make antibodies and killer T cells, the innate system must sense that there is danger. And one of the main clues that a dangerous viral attack is underway is the death of infected cells. Because we hepatitis B viruses usually do not kill the cells we infect, the innate system must wait for the odd infected liver cell to "make a fatal mistake" before it is alerted. Consequently, by blocking interferon production, and by not killing the liver cells I infect, I can "sneak up on" the immune system. This frequently allows me to establish a stronghold from which I cannot be dislodged—a feature that is essential for my lifestyle as a virus which can establish a long-term infection.

Eventually, however, virus-specific killer T cells and B cells are activated. These B cells produce antibodies that recognize the major viral antigen on the surface of my envelope, HBsAg. Normally, these antibodies would mark newly made viruses for destruction. However, my empty viral envelopes also carry this antigen on their surface, and there are so many of these empties that they "soak up" the relatively smaller number of anti-HBsAg antibodies. That's why I don't worry about making a lot of these empty virus particles. Production of empty envelope "decoys," which soak up neutralizing antibodies, is very effective in eluding the antibody defense. This places the burden of repelling my attack squarely on the shoulders of virus-specific killer T cells. In some cases, killer T cells are successful, and I must settle for an acute infection. However, many times I do "win," and establish a chronic, usually lifelong infection.

Because neutralizing antibodies are soaked up by empty envelope decoys, large amounts of infectious hepatitis B virus can remain in the circulation of chronically infected individuals. As uninfected liver cells proliferate to replace those killed by the immune response, these "fresh" liver cells can also be infected by the circulating virus, perpetuating the chronic infection. We start slowly, but hepatitis B–infected cells eventually produce so much virus that the blood of infected individuals (carriers) frequently contains about 100 million infectious particles per milliliter.

HOW DOES HEPATITIS B VIRUS SPREAD?

I: I suppose that explains why you are so infectious.

HBV: It certainly does. Hepatitis B virus ranks as one of the most infectious of all viruses: Transfer of a fraction of a drop of blood (as little as one microliter) is sufficient to spread the infection from one human to another. In fact, because carriers of hepatitis B usually have so much infectious virus in their blood, any scenario you can imagine in which blood or blood products are exchanged has a high probability of transmitting a hepatitis B infection.

I: That's rather frightening! So hepatitis B virus usually is spread by contaminated blood?

HBV: Actually, we viruses don't like the word "contaminated"—it has such negative connotations. We prefer to say that a hepatitis B infection occurs efficiently as a result of blood-to-blood contact. The liver is a large organ which is strategically positioned to intercept blood as it circulates through the body. In fact, about 25% of the total cardiac output passes through the liver with each beat of the heart. This makes the cells of the liver readily accessible to blood which contains virus particles.

I: So how does this "blood-to-blood contact" usually happen?

HBV: In its most natural setting, a hepatitis B infection takes place during the birth of a child—so-called "vertical" transmission. I say "most natural" because, although a hepatitis B infection can occur when drug addicts share needles, we certainly did not evolve to be spread in this way. That's nasty. And virus-containing blood transfusions are not our thing either, although we can be spread efficiently that way too. No, we evolved to infect humans via the "perinatal" route—by passing from the blood of an infected mother to the blood of her child during childbirth. Indeed, about 20% of babies born to hepatitis B–infected mothers will be infected at birth.

My goal is to produce a chronic infection with high levels of virus in the blood of my host. Because I spread from mother to child, an acute infection, which lasts only a short time, is useless to me. Moreover, humans are my only natural host. Consequently, to survive, I must persist in the body of an infected woman long enough to ensure that this "carrier" will be able to infect her children at birth. Fortunately, about 90% of infected newborns do end up as chronic carriers. These infected babies provide a relatively large pool of mothers who can infect their offspring, passing the infection from generation to generation.

I: Any other common routes of infection?

HBV: Another rather natural route of hepatitis B infection is from child to child, through open sores or cuts. We use this route of "horizontal" transmission when many children are crowded together, and hygiene is lax (e.g., in some daycare centers). Hepatitis B virus also is found in seminal fluid, and although sexual transmission is common, it is relatively inefficient. Unfortunately, for humans infected as adults, the outcome is much less favorable for us viruses. Indeed, in about 97% of adults, the infection is terminated, and damaged liver cells are replaced through the proliferation of healthy ones. You can see why we didn't evolve to infect adults. It would not have been worth the effort. These days, however, with humans abusing drugs and giving unscreened blood, adult infections are more promising. Things are looking up.

PATHOGENESIS

I: I know that your goal is to keep a low profile, but if you didn't cause disease in humans, I wouldn't be interviewing you.

HBV: That's true, I guess. But I could certainly do without the disease. It's really just an unintended consequence of a hepatitis B infection. In about 70% of infected adults, the immune response to the infection is so vigorous that the virus is eradicated with few or no symptoms. For the other 30%, destruction of liver cells by T cells is severe enough to cause the symptoms commonly associated with liver damage: elevated levels of liver enzymes (e.g., aminotransferases) in the serum, nausea, vomiting, liver pain, jaundice, and dark-colored urine. These symptoms can last for several months before the immune system subdues the infection.

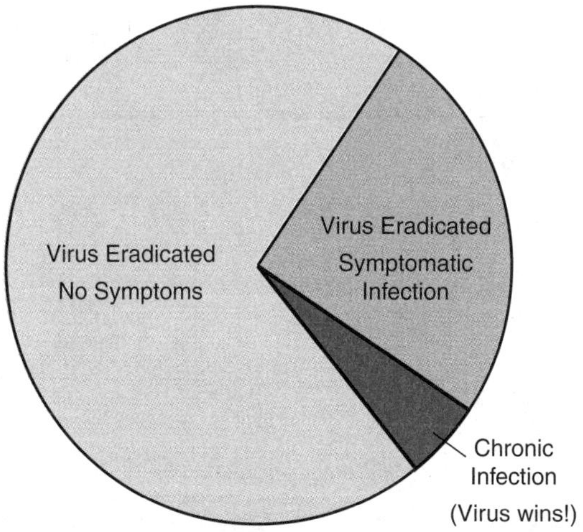

I: That certainly doesn't sound too pleasant. What about the people who are chronically infected?

HBV: Although the immune system of some chronically infected humans eventually does eradicate the virus, most chronic infections are lifelong. Many of these people remain relatively asymptomatic, whereas others suffer recurring bouts of liver inflammation which lead to more and more liver damage (cirrhosis). Also, complexes between the abundant decoy viruses in the blood and the antibodies that bind to them can lead to skin rashes, painful joints, and kidney disease when decoy-antibody complexes are deposited in these areas. However, I want to point out that hepatitis B virus usually does not kill the liver cells it infects, so most of the destruction of liver cells that occurs during a hepatitis B infection is the result of the immune response to the virus attack.

I: But what about liver cancer?

HBV: It's true that roughly 20% of long-term, hepatitis B carriers eventually contract liver cancer, and that about one million people die of hepatitis B–associated hepatocellular carcinoma each year. However, you must understand that hepatitis B virus does not "cause" cancer. All we "cancer viruses" do is increase the risk that an individual will accumulate the mutations required to turn a normal cell into a cancer cell. Anyway, it usually takes many years for cancer-causing mutations to accumulate, and hepatitis B–associated liver tumors generally arise 20 to 50 years post infection.

I: I can hardly wait!

THE INTERVIEWER'S SUMMARY

The most natural route of hepatitis B infection is from an infected mother to her baby during birth. However, this is one of the world's most infectious viruses, so any scenario by which blood from an infected individual is transferred to the bloodstream of another human can result in infection of its target cells—the hepatocytes of the liver. Hepatitis B is a DNA virus which replicates by a rather bizarre strategy, involving reverse transcription of an RNA intermediate to produce viral DNA. The result is a mostly double-stranded DNA genome which is enclosed in a protein capsid, and covered by an envelope derived from the endoplasmic reticulum.

Hepatitis B virus is non-cytolytic, so early in an infection, there is no cell death to help alert the immune system to the infection. In addition, this virus encodes proteins which effectively block the production of interferon. Later in infection, when anti-viral antibodies are made, they are diverted from their targets by a plethora of empty "decoy" virus particles. By using these evasion tactics, hepatitis B virus is able to establish a robust, acute infection during which huge amounts of virus are made by infected liver cells, and poured into the bloodstream of an infected individual. Although the immune system, led by killer T cells, eventually destroys the virus in most infected adults, in a small percentage of adults and in the majority of infected children, the virus establishes a chronic infection. This infection can last a lifetime, during which the infected individual continues to produce infectious virus, and the immune system does its best to keep the infection in check. A chronic hepatitis B infection can result in cirrhosis of the liver and liver cancer.

GENERAL PRINCIPLES

1. Viruses prefer to infect big organs (e.g., the liver), not only because such organs contain many target cells, but also because many cells can be infected without compromising organ function.
2. Viruses which are transmitted vertically from mother to child must establish a chronic infection in their host until she is old enough to bear children.

THOUGHT QUESTIONS

1. What features of hepatitis B replication are unique to this virus?
2. Why does the virus make so many empty envelopes?
3. What features of a hepatitis B infection are essential for ensuring that the virus can spread efficiently during a chronic infection?
4. Is it possible to become infected by hepatitis B virus by shaking hands with someone who is infected?

Chapter 11

Hepatitis C Virus: An Escape Artist

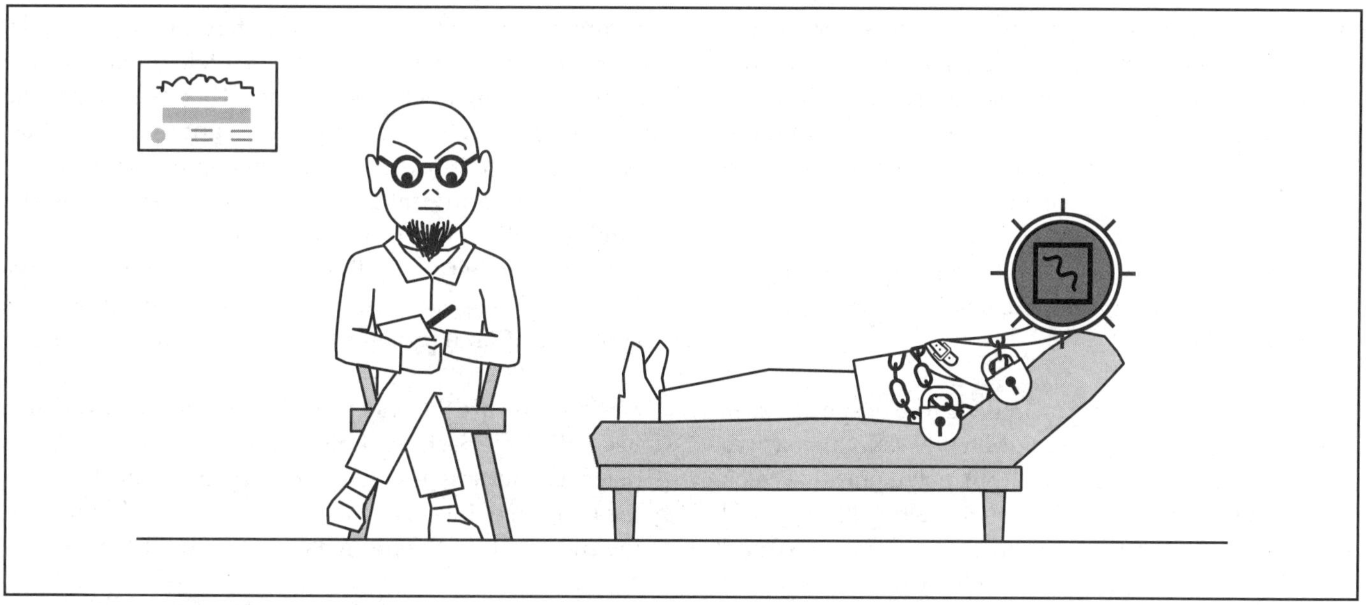

BACKGROUND

Hepatitis C is the third virus in our Bug Parade which targets liver cells. In the United States, about 60% of all newly diagnosed cases of chronic hepatitis are caused by hepatitis C infections, and worldwide, more than 350,000 people die each year from hepatitis C–associated liver disease. Amazingly, in the United States, about five times as many people are infected with hepatitis C as with HIV-1—with each virus killing more than 10,000 every year.

A HEPATITIS C VIRUS INFECTION

Interviewer: Your name implies that you infect liver cells, and I presume that you reach the liver through the blood—a hostile environment for most viruses. However, I've heard that you can be destroyed by soap and water.

Hepatitis C Virus: Quiet! Don't spread that around! I have an envelope which protects me from deadly enzymes present in the blood. My envelope even resists attacks by the complement system. However, my lipid coat dissolves when treated with soap and water. It's a little-known fact—and I'd like to keep it that way.

I: I'll bet you would. Because blood circulates throughout the body, I would think that being equipped to travel through the blood would give you a perfect opportunity to establish a systemic infection. So how do you manage to infect liver cells and not other cells in the body?

HCV: Oh, I'm very careful not to "go systemic." That would be counterproductive. I have evolved a very complicated system to gain entry into the cells I infect. I use a combination of four receptor proteins—one of which (SR-BI) is expressed almost exclusively on liver cells. That's how I stay focused on the liver.

VIRAL REPRODUCTION

I: Once you reach the liver, and bind to those receptor proteins, how do you reproduce?

HCV: My genome consists of a single strand of positive-sense RNA enclosed in a protein capsid surrounded by an envelope acquired from the infected cell. After my envelope fuses with the cell's plasma membrane, the capsid containing my genome is released into the cytoplasm. There, my viral RNA is translated in a cap-independent fashion using an internal ribosome entry site to yield a single long polyprotein.

I: Hepatitis A virus also encodes a polyprotein, which is subsequently cut up to make smaller proteins. Is that the way you operate?

HCV: Yes, my monster protein is cleaved at specific sites to yield 10 functional proteins. My viral RNA also serves as a template for my personal RNA polymerase—which is extremely error-prone—to make negative-strand RNA. My polymerase then copies this negative strand to make many genomic RNAs, which are encapsidated in protein, enveloped in a membrane derived from the cell's endoplasmic reticulum, and released from the cell.

VIRAL EVASION

I: My understanding is that you evolved to be transmitted from mother to child, so I presume that you are able to persist in the liver of your hosts until they are old enough to pass the virus along to their offspring.

HCV: Yes, hepatitis C can establish a lifelong, chronic infection. In fact, the majority of my hosts become chronically infected "carriers."

I: That kind of lifestyle must present problems in terms of evading host defenses—since you remain in the body for so long.

HCV: Yes, that was a difficult problem to solve. During hepatitis C replication, long stretches of double-stranded viral RNA are present, and this characteristic feature is detected by the TLR3 pattern recognition receptor. To keep this "alarm" from activating the interferon defense, I encode a two-protein complex (NS3/4A) which functions as a "dual-use" serine protease. I use this protease to cleave my polyprotein to produce functional viral proteins, but I also use NS3/4A to cleave one of the important proteins, TRIF, which TLR3 uses to signal. The result is that the TLR3 signaling cascade, which normally would turn on interferon-β production, is blocked before it reaches the nucleus. However, that block is not complete, so to deal with any interferon which is made, another of my proteins (the "core" protein) interrupts the signal from the interferon receptor that can induce the expression of cellular proteins with antiviral activities. And just in case this signal is not completely blocked, and cellular antiviral proteins are synthesized by the infected cell, I encode a protein, NS5A, which inhibits the kinase activity of PKR, and blocks the antiviral function of 2′-5′ oligoadenylate synthetase. This allows the synthesis of viral proteins to continue in infected cells.

I: Wow! You are the only virus I have interviewed which uses all three tactics to protect against the interferon defense: interfering with interferon production, blocking the turn-on of interferon-inducible genes, and diminishing the effects of the proteins encoded by these ISGs.

HCV: That's one reason I can establish a lifelong, chronic infection. I'm really careful about interferon. But don't be too sure about those other viruses. They are using more schemes for protection from interferon than they admit. Virologists don't know the half of it, and we viruses certainly are not going to tell them the whole story!

I: Are you at liberty to tell us how you manage to evade the adaptive immune system?

HCV: Yes, of course. That's not a big secret. Within an infected individual, I can be spread when newly made viruses leave one cell and infect another. However, hepatitis C virus also can be spread within its hosts by cell-to-cell contact. By creeping from one cell to another, I avoid the antiviral antibodies that can neutralize viruses which are outside of cells. Of course, this type of spread cannot evade killer T cells. Those awful cells are the ones we really have to watch out for as we attempt to establish a chronic infection.

I: So how do you deal with killer T cells?

HCV: I rely on my error-prone, RNA polymerase to introduce many mutations into my genome. These mutations produce a large number of genetically different, but closely related, "quasispecies," and this allows me to "escape" from the host's immune defenses. But it's a real roller coaster ride! During the initial infection, antibodies and killer T cells are produced which recognize the infecting virus. Then, just when the immune system has almost wiped us out, mutations in my genome give rise to one or more new variants which cannot be recognized by the original killer T cells or antibodies. As a result, the immune system must start from scratch to produce different B cells and killer T cells—ones which are appropriate to defend against the mutated virus. And while this is going on, we are reproducing and creating more quasispecies to try to keep one step ahead. You might call us "escape artists." It's quite nerve-racking, but that's what we must do to maintain a chronic infection. Sometimes we win, and sometimes we lose. But that's how we play the game.

HOW DOES HEPATITIS C VIRUS SPREAD?

I: Hepatitis C virus was first identified in 1989. Are you really a "new" virus, or is it just that advances in technology have only recently made it possible to detect a hepatitis C infection?

HCV: Our early family history is a bit hazy, and I can only tell you what has been passed down through the viral equivalent of oral tradition. The story goes that hepatitis C was originally a chimpanzee virus. One day, while a woman was preparing chimpanzee for dinner, she got a little careless with the knife, and cut herself. Some of the blood from the infected chimpanzee entered her blood stream, and the rest is history. We call these early hepatitis C viruses the Pioneers, because life was tough for them. These original viruses could be transmitted "vertically" by an infected mother to her newborn child when blood was exchanged during the trauma of childbirth. But vertical transmission is not very efficient. Probably less than 10% of infected mothers pass the virus on to their children. Also, babies are not infected during breastfeeding, and although hepatitis C virus is found in saliva and semen, transmission by sexual contact is inefficient. There was also some horizontal transmission when infected children "exchanged" blood with other children during rough play. Oh, and then there was the occasional, untalented barber who used the same dirty razor to shave both infected and uninfected men—or the surgeon or dentist who wasn't careful to sterilize his instruments.

I: So if hepatitis C infection is that inefficient, and transmission from chimpanzees to humans is a rare event, I can't imagine how you manage to persist in the human population.

HCV: Yes, the Early Years were tough going. But then things began to improve. Probably the first bright spot was the advent of the widespread use of blood transfusions in the 1940s—transfusions which had not been screened for the presence of hepatitis C virus. That was really a boon, because this practice made transmission from human to human much easier. Then, about the same time, the number of people who inject drugs for "recreational purposes" began to increase. These folks generally are so wacked out that they share needles and other paraphernalia without even washing them! I love those addicts.

Also, there were vaccination programs with lax hygiene. These were very helpful to us. Indeed, when hepatitis C viruses gather, the story is often told about the World Health Organization's campaign from the 1950s until the 1980s in Egypt. Because of inadequate sterilization and the reuse of glass syringes, hepatitis C virus was spread widely when millions of people there were treated for parasitic worms by intravenous injections. As a result of this windfall, one in seven Egyptians is currently infected.

Today, most people contract hepatitis C virus by receiving a transfusion of unscreened blood, by sharing needles during recreational drug use, or by the reuse of instruments during medical procedures. We young hepatitis C viruses owe a great debt to the Pioneers, because they paved the way for us. Indeed, I feel fortunate to be living in the "Golden Age of Hepatitis C Infection." After all, nearly 200 million people are now infected worldwide with hepatitis C virus—and most of them don't even know it. We've come a long way, baby!

PATHOLOGICAL CONSEQUENCES OF A HEPATITIS C INFECTION

I: You raise an interesting point. At least two-thirds of all new hepatitis C infections go unrecognized because they are asymptomatic or only mildly symptomatic. In about 25% of those infected, the immune system can fight off the infection and banish the virus from their liver. In the rest, hepatitis C virus establishes a smoldering, chronic infection that usually is not detected until 10 or more years after the initial exposure.

HCV: Yes, it's unfortunate that the immune system keeps us from achieving a perfect record of chronic infections. We keep trying, but it's a real battle.

I: What are the consequences of this battle for those who are chronically infected?

HCV: In about 20% of chronic infections, the continuing waves of infection which occur—as we mutate and escape from immune surveillance—eventually lead to cirrhosis of the liver. This generally occurs about two decades post infection when the killing of liver cells by the immune response produces scar tissue which disrupts the architecture and function of the liver.

I: I suppose you would contend that, because it is the immune response which kills those liver cells, you are not to blame for the pathological consequences of a hepatitis C infection.

HCV: I'm not sure I would go quite that far. I will accept some of the blame, but cirrhosis certainly is an unintended consequence of a hepatitis C infection. I'm a non-cytolytic virus, and all that cell killing and scarring doesn't help me at all. Moreover, hepatitis C infections and excessive alcohol consumption can act synergistically to accelerate the progression from infection to cirrhosis. And you certainly can't blame me for making those folks drink.

I: No, I guess not. But what about liver cancer?

HCV: About 10% of patients with hepatitis C–induced cirrhosis of the liver eventually suffer from liver cancer (hepatocellular carcinoma), which usually arises about three decades post infection. But that's clearly another unintended consequence. After all, a dead host is not a good host.

I: I think I've heard that one before.

THE INTERVIEWER'S SUMMARY

Hepatitis C virus is a single-stranded, positive-sense RNA virus whose genome is protected by a protein capsid and a cell-derived envelope. It infects liver cells, but does not kill them. The virus uses a complicated set of cellular receptors to focus on its target cell, and after the viral envelope fuses with the cell membrane, the genome-containing capsid is released into the cytoplasm. There the viral RNA is translated using an internal ribosome entry site to synthesize a long polyprotein—which is cleaved to yield the proteins the virus needs to reproduce.

This virus can cause either an acute or a chronic infection, and it uses three different strategies to evade the interferon defense: It partially blocks interferon production, it interferes with the expression of interferon stimulated genes, and it compromises the function of the protein products of these ISGs. To stay one step ahead of the adaptive immune system, the virus uses its error-prone RNA polymerase to introduce mutations that result in the production of proteins the immune system has not seen before. In addition, hepatitis C virus can evade antiviral antibodies by spreading within the host by cell-to-cell contact. This virus can be transmitted vertically from mother to child during birth or by any other route which involves the exchange of blood or blood products.

GENERAL PRINCIPLES

1. Only viruses which cause long-term infections are associated with cancer in humans.
2. Frequently, the pathologic consequences of a viral infection are unintended, and are not advantageous to the virus.

THOUGHT QUESTIONS

1. Hepatitis C virus and rhinovirus employ similar replication strategies. What are three ways in which these viruses are different?
2. How does hepatitis C virus manage to "escape" from the immune defenses?
3. Why would it be counterproductive for hepatitis C virus to "go systemic"?

Chapter 12

HTLV-I: A Tribal Virus

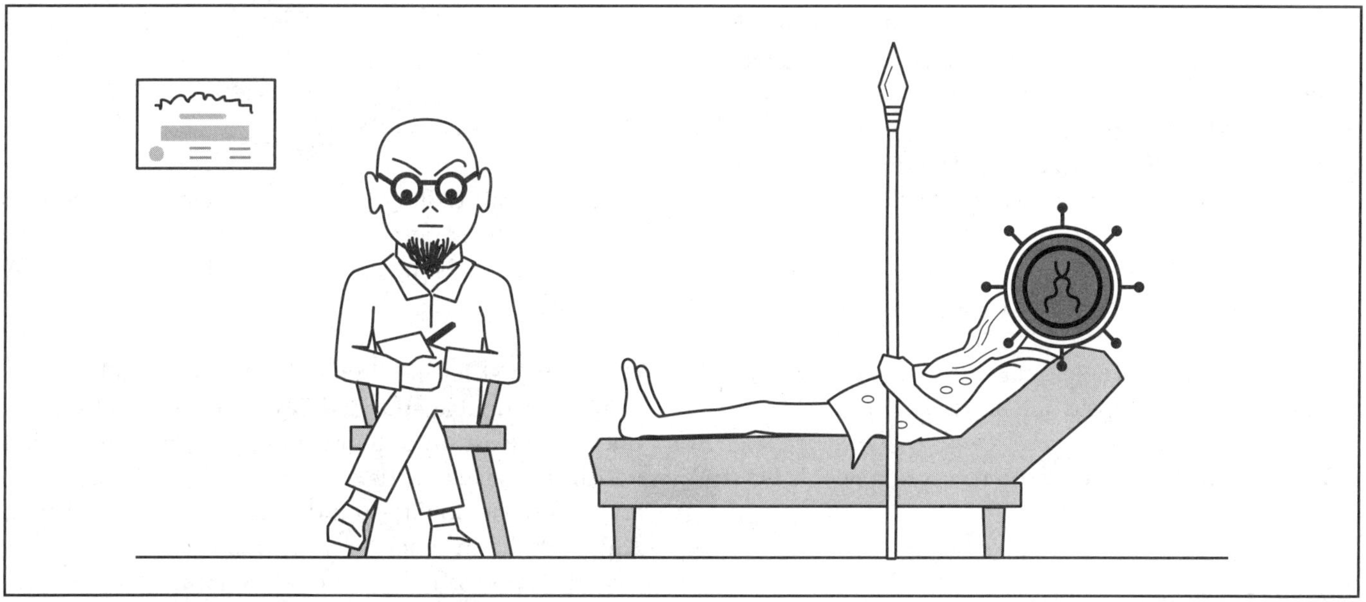

BACKGROUND

Retroviruses—viruses which have RNA genomes, but which replicate through a DNA intermediate—were first isolated from animals at the beginning of the nineteenth century. However, it was not until 1980 that the first human retrovirus, human T cell lymphotropic virus type I (HTLV-I) was discovered. Currently, it is estimated that about 20 million people worldwide are infected with HTLV-I. Since its discovery, other human retroviruses have been identified, and HTLV-I (a.k.a. human T cell leukemia virus) has been overshadowed by its more famous cousin, the human immunodeficiency virus (HIV-1)—which causes AIDS (originally called HTLV-III).

One of the fascinating features of HTLV-I infections is that they occur most frequently in geographically isolated populations (e.g., small villages on Japanese islands). Although virologists are still not sure why high rates of HTLV-I infection are found in isolated populations, it is possible that there is something in the genetic makeup of people in these small "tribes" which makes it easier for the virus to spread among them.

AN HTLV-I INFECTION

I: I'm told that you are a very old virus.

HTLV-I: Eh? Speak up, Sonny. I'm a little hard of hearing.

I: YOU DO LOOK LIKE A VERY OLD VIRUS!

HTLV-I: You don't have to shout! I'm old, but I'm not deaf. Yes, my family has been around for about 100,000 years. That's when our ancestor made the jump from a non-human primate to a human—at a time when most humans lived in small, isolated tribes.

I: Very interesting. What types of cells do you usually infect?

HTLV-I: I target cells of the immune system, primarily helper T cells and dendritic cells. To enter these cells, I use three receptor proteins on the cell surface. Two of them, NRP-1 and HSPG, are involved in the binding of my envelope to the target cell. The third, GLUT1, is required for fusion of my envelope with the cell's plasma membrane. Once my envelope has fused with the plasma membrane, the two copies of my single-stranded RNA genome are released into the cytoplasm of the cell.

I: So HTLV-I has a single-stranded RNA genome, protected by a protein capsid and an envelope.

HTLV-I: Didn't I just say that? You young people don't listen very carefully.

VIRAL REPRODUCTION

I: What happens after you reach the cytoplasm?

HTLV-I: Just what you'd expect: My capsid is removed, and a viral enzyme (reverse transcriptase), which is packaged inside my capsid, goes to work. This enzyme copies my single-stranded RNA genome to produce a single-stranded, complementary DNA (cDNA) molecule and destroys the original RNA molecule after it has been copied. The reverse transcriptase protein then makes a complementary copy of the single cDNA strand to produce a double-stranded cDNA molecule. The overall result of this reverse transcription is to replace my single-stranded RNA genome with a double-stranded DNA "copy" which contains my viral genetic information.

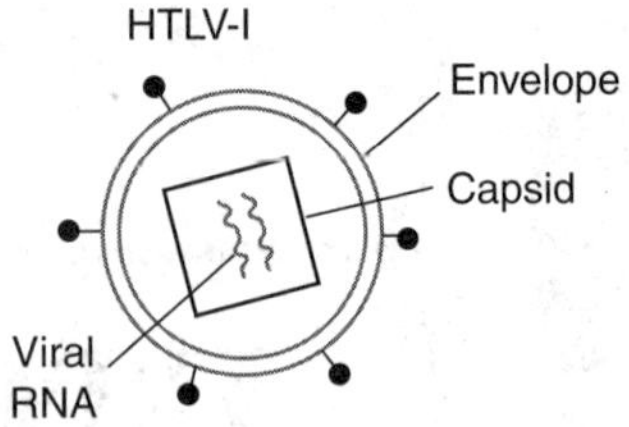

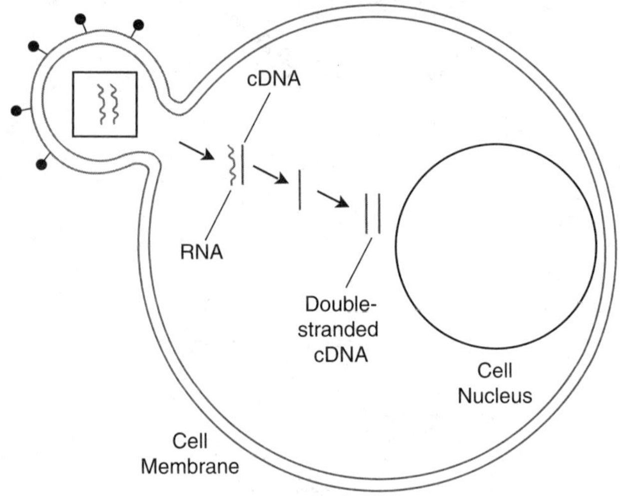

I: That certainly is <u>not</u> what I expected! Why would you switch from being a virus with a single-stranded RNA genome to being one with a double-stranded DNA genome?

HTLV-I: Just let me finish, Sonny, and you'll understand! The double-stranded DNA copy enters the cell's nucleus where my integrase enzyme, also packaged within my virus particle, helps cut the cellular DNA, and insert the double-stranded viral DNA at this spot. Although the chromosomal site into which the cDNA is inserted is essentially random, my viral cDNA is carefully integrated so that none of the viral genes is interrupted. Virologists call the viral cDNA which is inserted into the cell's DNA the "**provirus.**"

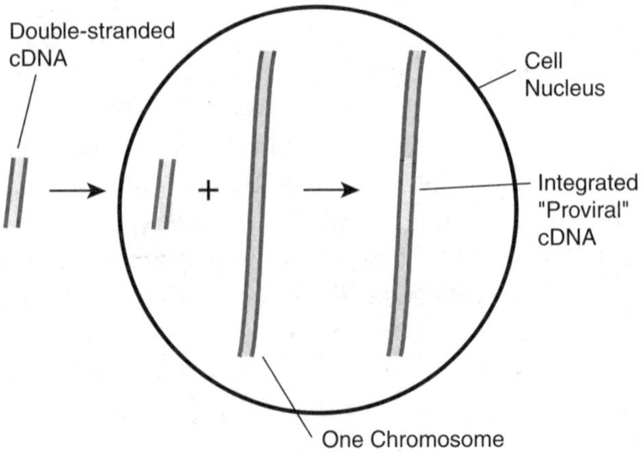

I: That all seems rather complicated.

HTLV-I: It's actually very simple. Of course, I've been doing it for 100,000 years, so I've had plenty of practice. But back to your question: You asked why I make this change from RNA to DNA, and here's the answer. HTLV-I's ability to actively integrate its genetic information into the target cell's DNA is key to producing a lifelong infection. Once integrated into the cellular genome, the HTLV-I sequences will be copied and passed down to daughter cells as the infected cell replicates its DNA and divides. Interestingly, integration of HTLV-I proviral DNA cannot take place unless the target cell is in the process of replicating its DNA. This means that resting cells (most of the cells in the body) are not good targets for HTLV-I infection.

I: Doesn't this severely limit the cells you can infect?

HTLV-I: That depends on the definition of the word "infect." It certainly does make the initial infection difficult. However, once those first cells are infected, I'm in business. In addition to viral genes that encode my reverse transcriptase enzyme and my coat proteins, my genome also encodes six regulatory proteins, including one called Tax. This protein is produced early after infection, and it modulates the expression or function of cellular genes, and causes HTLV-I–infected cells to proliferate. And when these infected cells proliferate, the integrated provirus is copied right along with the rest of the cell's DNA, and is passed down to each of the daughter cells—resulting in two infected cells. Next, the Tax protein produced in these daughter cells causes them to proliferate, resulting in four infected cells. And so it continues. Consequently, by first integrating my genetic information in the form of a provirus into the genome of my target cell, and then by driving this cell to proliferate, I am able to "infect" a large fraction of a person's helper T cells without needing to produce infectious virus.

I: Very clever. But aren't you stuck there once the provirus is integrated?

HTLV-I: No, of course not. That's just the first step—a very important step, because no HTLV-I mRNAs or genomes can be produced until the provirus has been inserted into a cellular chromosome. However, once that has been accomplished, my genetic information can be transcribed by the cell's RNA polymerase to make either short mRNAs—which encode my viral proteins—or full-length RNAs that will be used as new viral genomes. Eventually, two of these genomic RNAs are packaged into a protein capsid. Then, the newly made virus leaves the cell by budding through the plasma membrane, picking up an envelope which contains both viral and cellular proteins. Importantly, HTLV-I–infected cells usually are not killed in the process of producing new viruses.

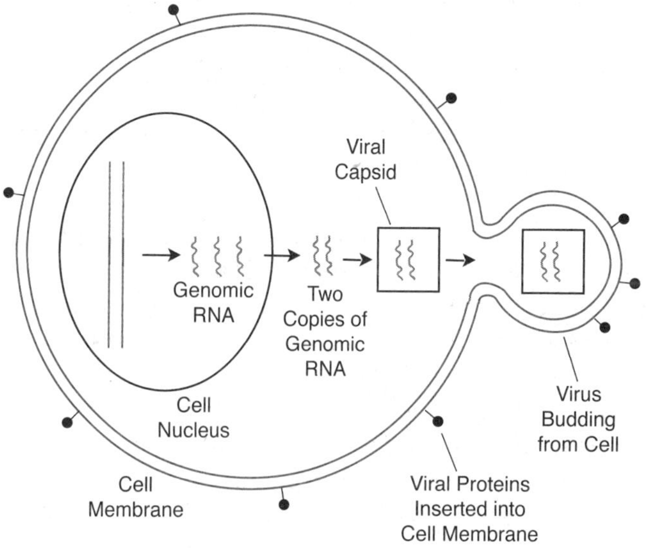

I: Pretty slick.

HTLV-I: Thanks, but we have had a long time to work out this strategy. Did I tell you I was a very old virus?

I: Yes, I think you did.

HOW DOES HTLV-I EVADE HOST DEFENSES?

I: Since you have been around for so long, I'm sure you've evolved ways of evading host defenses.

HTLV-I: Indeed I have. My most effective evasion tactic is my ability to establish a stealth (latent) infection in which the integrated provirus is passed down to daughter cells when infected cells proliferate. Most of the time when this happens, little or no viral RNA is transcribed from the integrated, proviral genome. Consequently, to the immune system, these infected cells are almost indistinguishable from uninfected cells. That's a good thing, because humans infected with HTLV-I eventually produce large numbers of killer T cells that can recognize my viral proteins.

I: Because you can "multiply" within your infected host simply by riding along in the cellular genome, it would seem that you are safe from the adaptive immune system.

HTLV-I: Actually I'm not. There are two reasons. First, most of the immune system cells I infect normally would not continue to proliferate. And if they don't proliferate, I can't use this method of spread. So to increase the number of infected helper T cells, I am forced to occasionally produce Tax proteins which cause these infected cells to proliferate. In addition, to produce new viruses, my integrated provirus must be transcribed, and viral proteins must be synthesized.

I: So you must walk a fine line between hiding from killer T cells, which can destroy HTLV-I–infected cells, and producing viral proteins which can either drive cell proliferation or be used to construct new virions.

HTLV-I: That's true, but fortunately I have a way of reducing the risk of detection by killer T cells. These assassins must view viral proteins presented on the cell surface by class I MHC molecules in order to know that a cell has been infected. To make detection difficult for them, I encode a protein, p12, which interferes with the presentation of viral proteins by class I MHC molecules.

I: Because you are an RNA virus that replicates using an error-prone reverse transcriptase, I would think that you also could use antigenic drift to evade the adaptive immune response.

HTLV-I: That would be true if I "reproduced" mainly by repeated rounds of viral replication using my error-prone polymerase. However, HTLV-I usually spreads within an infected individual when cells containing an integrated provirus proliferate, not by repeated rounds of viral infection. And during cell proliferation, my integrated provirus is copied by the high-fidelity DNA replication machinery of the cell. As a result, antigenic drift isn't an effective evasion mechanism.

I: Oh, I almost forgot to ask: Do you have ways of defending yourself against the interferon system?

HTLV-I: Of course I do! You don't get to be 100,000 years old without being able to defend yourself against interferon. In fact, my defenses against that system are so time-tested, they are "classified."

HOW HTLV-I SPREADS

I: In addition to the proliferation of cells with integrated provirus, are there other ways you spread within an infected individual?

HTLV-I: Yes, we actually have three ways of spreading within our hosts during an infection. The most important way is by proliferation of cells with integrated provirus. In fact, with established HTLV-I carriers, roughly 1% of all helper T cells will have an integrated provirus. That's about one billion cells! But HTLV-I also can spread when viral genomic RNA (complexed with several viral proteins) is transferred from an infected cell to an uninfected cell by cell-to-cell contact. During this transfer, the virus never leaves the protected environment of a human cell. That's advantageous, because viruses that are traveling through the body from one cell to another are fair game for antiviral antibodies which can tag them for destruction.

In addition to these two "virus-less" strategies, when proviruses in latently infected cells "reactivate" and begin to produce new virus particles, these new viruses can be exported out of the cell and infect other immune system cells. However, despite the large number of infected cells, HTLV-I carriers usually have almost undetectable amounts of virus circulating in their blood. One of the reasons for this is that the number of viruses produced by an HTLV-I–infected cell is rather small. Moreover, the virus particles produced by HTLV-I–infected cells actually are not very infectious—only about one in 100,000 of all viral particles is competent to produce an infection. That's why the proliferation of latently infected cells is the primary way we spread within infected humans.

I: But if you produce only a small amount of infectious virus, how are you transmitted from person to person?

HTLV-I: Infections spread most efficiently when HTLV-I–infected cells, not virus particles, are transferred from one human to another. Infected cells are present in blood, in breastmilk, in semen, and to a much lesser extent, in vaginal secretions of persons who have been infected with HTLV-I. One of the most "ancient" modes of transmission is breastfeeding, in which infected cells are transferred from a mother to her child. In fact, the likelihood of an infected mother spreading the infection to her baby by breastfeeding is roughly 20%, with the probability of spread increasing with longer periods of breastfeeding. The virus produced by the infected cells which these babies receive with mom's breastmilk can establish a chronic infection that lasts for life. Consequently, once these infected babies reach puberty, they can infect their mates during sexual intercourse or pass the virus to their offspring during breastfeeding. By establishing a chronic infection in which most viruses are latent, we give an infected person a whole lifetime of chances to "pass it on."

I: You said breastfeeding was an ancient way HTLV-I spreads, so I suspect there are more "modern" ways. For example, can people be infected with HTLV-I when they receive blood products like clotting factors?

HTLV-I: Unfortunately, blood products rarely transmit HTLV-I infections—because they don't contain infected cells. However, infected cells in unscreened whole blood can transmit an HTLV-I infection. In addition, two new susceptible populations have emerged: intravenous drug abusers and persons infected with the AIDS virus. Intravenous drug abusers can spread HTLV-I–infected cells very efficiently by sharing contaminated needles, and many persons infected with HIV-1 engage in unprotected, promiscuous sex, increasing the number of people that a single infected individual is likely to infect. In addition, members of both groups tend to be immunosuppressed—either by virtue of an HIV-1 infection or because of a generally unhealthy lifestyle—and therefore are more susceptible to HTLV-I infection.

VIRAL PATHOGENESIS

I: What are the consequences of an HTLV-I infection?

HTLV-I: Our genetic information is "silent" in the overwhelming majority of infected cells, and we do not kill the cells we infect. In fact, the initial infection with HTLV-I is uniformly asymptomatic. Most people never even know that they are infected. However, because so many immune system cells eventually are infected, especially helper T cells, my hosts do become more susceptible to the same types of opportunistic infections which are experienced by people infected with HIV-1.

I: And aren't you called a "leukemia virus"? Were you incorrectly named?

HTLV-I: No. About 2% of those infected with HTLV-I eventually contract adult T cell leukemia (ATL). In addition, HTLV-I–infected humans have about a 2% lifetime risk of suffering from a neurological disease with a very long name: HTLV-I–associated, myelopathy/tropical spastic paraparesis (usually called HAM/TSP). This is a progressive, paralytic disease in which the long motor neurons in the spinal cord become demyelinated. Like adult T cell leukemia, HAM/TSP usually is diagnosed when patients have been infected for 40 or 50 years. Of course, when a person's immune system has been battling a smoldering viral infection for that long, you might expect something to go wrong.

I: HAM/TSP doesn't sound like a good disease to have. What happens to the people with adult T cell leukemia?

HTLV-I: It's really sad, but once this cancer is diagnosed, the victim usually dies within a year. In the Old Days, this wasn't a problem. Back then, we HTLV-I viruses were totally benign, just minding our business, and being passed from mother to child in breastmilk or during the joys of sex. In the ancient tribes which we evolved to infect, people simply didn't live long enough for us to cause these problems: Humans usually died of infectious diseases or injuries long before any HTLV-I–associated disease would bother them. In fact, it's only recently that we got the name "leukemia virus."

I: I guess you've outlived your times. You do look awfully old.

HTLV-I: Yes, and all this talk has really tired me out. I think I'll take a nap.

THE INTERVIEWER'S SUMMARY

HTLV-I is a very old virus that evolved to be passed from an infected mother to her child during breastfeeding (vertical transmission) or from one adult to another during sex. This virus, which infects helper T cells and dendritic cells, has a single-stranded RNA genome, protected by a protein capsid and a cell-derived envelope. Upon infection, this retrovirus uses its reverse transcriptase enzyme to convert its single-stranded RNA into double-stranded DNA, and then uses another viral enzyme, the integrase, to integrate this DNA into a cellular chromosome. Both enzymes are packaged within the virus particle, so they are ready to go to work as soon as the virus enters the cell.

Once integrated into the cell's chromosome, the "proviral" DNA can be passed down to daughter cells as infected cells proliferate. In its integrated state, few or no viral proteins are produced, and to help avoid detection by killer T cells, HTLV-I encodes a protein which interferes with the display of viral proteins by class I MHC molecules. In addition, genomic RNA can be transferred from infected to uninfected cells by cell-to-cell contact, and this helps the virus avoid being tagged for destruction by antiviral antibodies during spread within an infected host.

HTLV-I usually is transmitted from human to human during the transfer of virus-infected cells, not virus particles. In addition to the "natural" avenues for viral transmission—breastfeeding and sex—HTLV-I–infected cells also can be transmitted by contaminated needles or by blood transfusions. HTLV-I does not kill the cells it infects, and the systemic infection is largely asymptomatic. This ability to live "happily" with its infected host for long periods is probably a result of many thousands of years of co-evolution.

GENERAL PRINCIPLES

1. When they infect cells, retroviruses produce DNA proviruses that can be integrated into the cell's DNA. There they can remain "silent" as they are passed down with the cell's chromosomes when the cell proliferates.
2. New viral genomes and viral mRNA are not synthesized until after the proviral DNA of a retrovirus has been integrated into a cell's chromosome. Consequently, the viral enzymes required to produce the proviral DNA and integrate it must be packaged with the viral genome.

THOUGHT QUESTIONS

1. Which features of its replication style define a retrovirus?
2. What are the advantages of having the provirus integrated into the cell's DNA?
3. Why does HTLV-I have a low overall mutation rate even though its reverse transcriptase is extremely error-prone?
4. Why are retroviruses not designated as positive- or negative-strand viruses?

A Comparison of Vertically Transmitted Viruses

Let's review some important features of the three viruses we discussed which can be transmitted "vertically" from mother to child: hepatitis B virus, hepatitis C virus, and HTLV-I. Hepatitis B and C viruses target cells of the liver (hepatocytes), whereas HTLV-I infects immune system cells (especially helper T cells). The genome of each of these viruses is protected by a protein capsid enclosed in a cell-derived lipid envelope. During infection, this envelope fuses with the target cell's plasma membrane, releasing the partially encapsidated genome into the cell's cytoplasm. HTLV-I is a retrovirus which employs its reverse transcriptase enzyme to produce a double-stranded DNA "copy" (the provirus) of its single-stranded RNA genome. This proviral DNA is then carefully inserted into one of the host cell's chromosomes through the action of another enzyme, the viral integrase.

Hepatitis B is a mostly double-stranded DNA virus which also encodes a reverse transcriptase. This enzyme is used in a rather bizarre replication scheme to produce new viral genomes. However, because hepatitis B virus lacks an integrase enzyme, the DNA of hepatitis B virus is not integrated into host cell chromosomes. Hepatitis C virus has a single-stranded, positive-sense RNA genome. This RNA is translated in a cap-independent manner, making use of an internal ribosome entry site. The result is a single long protein which is subsequently cleaved to produce the viral proteins required for reproduction. All three viruses are non-cytolytic.

Hepatitis B and C viruses both evade the interferon defense by blocking interferon production. Hepatitis C virus also produces proteins which interfere with the expression of interferon stimulated genes, and which block the action of antiviral proteins encoded by these genes. Hepatitis B "confuses" the antibody defense by producing many empty "decoy" viruses, whereas hepatitis C virus uses its error-prone RNA polymerase to escape immune surveillance by staying one step ahead of the adaptive immune system. HTLV-I evades immune defenses primarily by establishing a latent infection in which few or no viral proteins are produced. And when viral proteins are synthesized, this virus compromises the function of killer T cells by interfering with the display of viral proteins by class I MHC molecules.

After the initial infection, hepatitis B and C viruses spread within their infected hosts when newly made viruses infect additional liver cells. HTLV-I integrates its genetic information into the genome of infected helper T cells, and when these cells proliferate, the integrated provirus is copied and passed down to the daughter cells—increasing the number of infected cells. HTLV-I and hepatitis C viruses also can spread "internally" during contact between infected and uninfected cells. All three viruses can cause lifelong, chronic infections.

Vertical transmission of hepatitis B and C viruses occurs mainly by the transfer, during the trauma of childbirth, of blood that contains infectious virus. In contrast, HTLV-I usually is transmitted vertically by the transfer of virus-infected cells during breastfeeding. The reason for the difference in avenues of vertical infection is that hepatitis B and C viruses both target cells of the liver—which are accessed via the blood. In contrast, HTLV-I infects helper T cells and dendritic cells which are accessible in the digestive tract, probably via the tonsils. Because the blood of hepatitis B carriers is chock full of virus, and because any blood transferred to another human almost immediately comes in contact with liver cells, hepatitis B is one of the most infectious viruses known. Hepatitis C virus and HTLV-I also can be transmitted horizontally by blood-to-blood contact, although HTLV-I is not very infectious, relying mainly on the transfer of infected cells to spread from human to human. All three viruses can be transmitted during sexual intercourse, but for hepatitis B and C viruses, this route of infection is relatively inefficient.

In part because they do not kill the cells they target, infection by these viruses is usually asymptomatic or only mildly symptomatic. When moderate to severe liver damage does occur, it is the result of the immune system's response to the hepatitis B or C infection. All three viruses are associated with human cancers that arise in the cell types they infect. Persons infected with HTLV-I have about a 2% lifetime risk of contracting adult T cell leukemia, and infection with hepatitis B or hepatitis C virus increases the risk for liver cancer. As is typical of virus-associated cancers, these malignancies usually arise decades post infection.

Chapter 13

VIRUSES WE GET BY INTIMATE PHYSICAL CONTACT

HIV-1: An Urban Virus

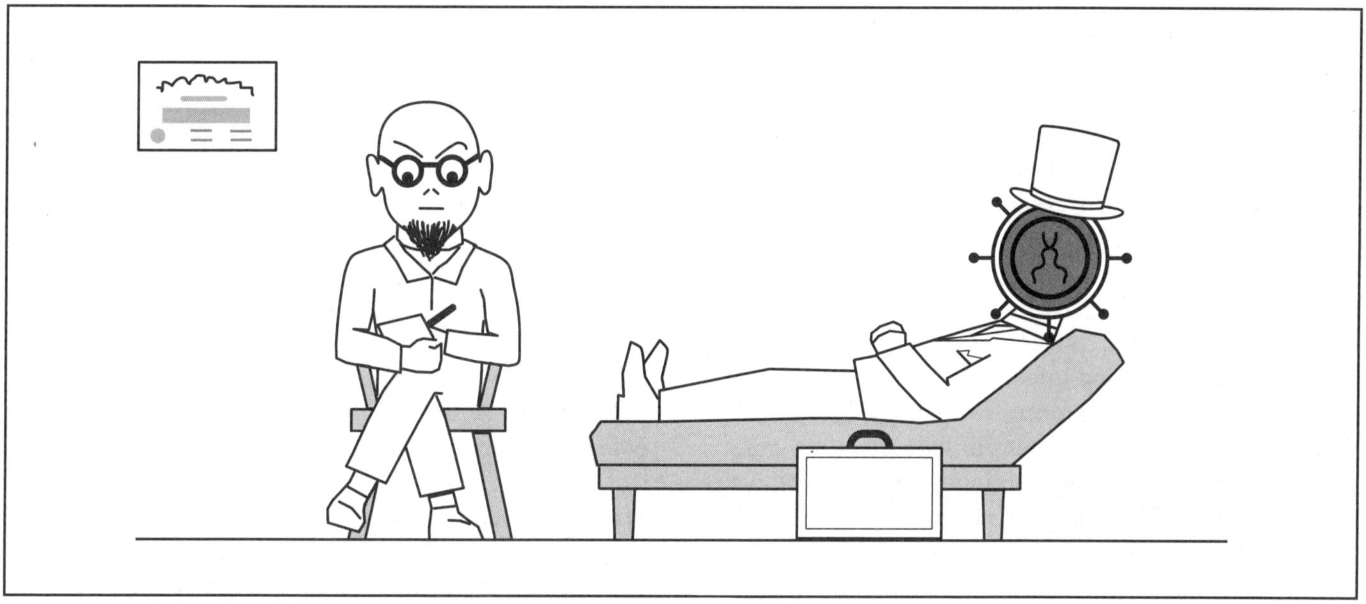

BACKGROUND

So far, we have interviewed viruses we inhale, viruses we eat, and viruses we get from mom. In each case, these viruses can be spread by mechanisms over which we have little or no control. For example, unless you never go out in public, at some time or other you are certain to have someone sneeze nearby and infect you with one of the respiratory viruses. In contrast, this chapter will focus on three viruses (HIV-1, herpes simplex virus, and the human papillomavirus) which, in most cases, can be avoided because they usually are spread by intimate physical contact. The fact that these viruses thrive in human populations is a demonstration of how strong the urge for intimate contact is—and how clever these viruses are to have evolved to take advantage of these urges.

Human immunodeficiency virus type one (HIV-1) is a close relative of HTLV-I. However, these viruses have very different lifestyles, which result in stunning differences in the pathological conditions they cause. HIV-1 likely "jumped" from infected chimpanzees to humans in Africa several times around the middle of the twentieth century—probably when blood from an infected chimpanzee was transferred to the bloodstream of humans during preparation of "bush meat." The chimpanzee ancestor of HIV-1 is thought to be a very old virus that evolved long ago to infect non-human primates without causing damaging symptoms. Because humans didn't just begin butchering chimpanzees in the middle of the twentieth century, it's probable that humans had been infected with HIV-1 on many earlier occasions—but that the virus just didn't "catch on" in the human population until very recently. Today, nearly 40 million people worldwide are infected with HIV-1.

HOW DOES HIV-1 INFECT ITS HOSTS?

Interviewer: When you infect humans, what cells do you target?

HIV-1: I specialize in infecting cells of the immune system that have CD4 proteins on their surface: helper T cells, macrophages, and dendritic cells.

I: So you infect the same cell types as does the human T cell leukemia virus I just interviewed.

HIV-1: Yes, we are close relatives. However, he has been around in the human population a lot longer than I have, and there are major differences in our lifestyles. For example, the most natural way for HTLV-I to reach these target cells is during breastfeeding. In contrast, my most natural way of reaching my targets is during sex, when virus or virus-infected cells breach the epithelial barrier of the vagina or anus and infect $CD4^+$ cells in the submucosa. These infected cells form a focus of infection which then spreads to other $CD4^+$ cells in the immediate area. This local infection usually continues for about 10 days. My goal during this time is to make my way, either on my own, or as "cargo" aboard dendritic cells, to nearby lymph nodes where many helper T cells are gathered. If I can infect lymph node cells, then newly made viruses can be carried by the lymph and blood throughout the body to cause a systemic infection of $CD4^+$ cells.

I: And these $CD4^+$ cells, how do you enter them?

HIV-1: Like HTLV-I, my genome is protected by a protein capsid and enclosed in a cell-derived envelope. Sticking out of this envelope is a protein, SU, which binds to the CD4 protein on the cell surface. This triggers a conformational change which then permits binding of my envelope to a co-receptor molecule. Indeed, CD4 binding by HIV-1 is not enough for viral entry: My envelope proteins must also bind to co-receptor molecules on the surface of my target cell. The co-receptor proteins I use most frequently are CXCR4 and CCR5, and binding to one of these co-receptors facilitates efficient fusion of my viral envelope with the cell's plasma membrane, allowing me to enter the cell's cytoplasm.

VIRAL REPRODUCTION

HIV-1: Because HTLV-I and I both are retroviruses, the strategies we use to replicate our single-stranded RNA genomes are very similar.

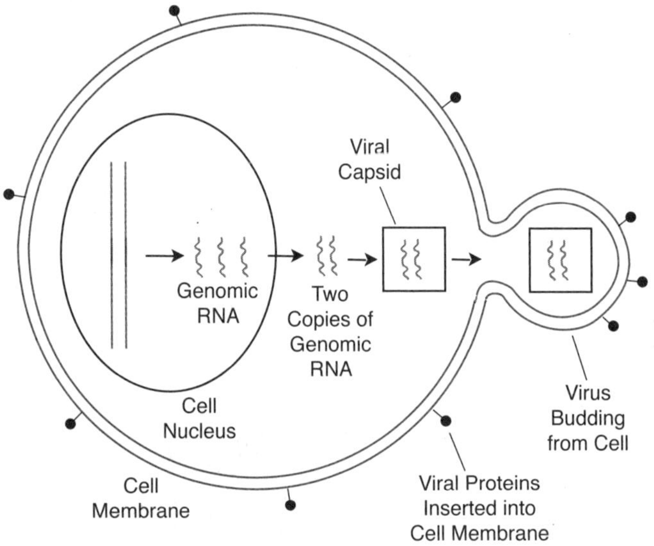

I: Yes, I can see the similarities, but there must be some subtle differences.

HIV-1: Indeed there are. We both use the cellular RNA polymerase to transcribe the integrated proviral DNA to synthesize our mRNA and our new viral genomes, and we both encode a number of accessory proteins. However, one of my accessory proteins, Vpr, makes it possible for me to integrate my proviral DNA into the chromosomes of some cells (e.g., macrophages) which are not proliferating. In contrast, HTLV-I is limited to infecting proliferating cells. Another important difference is that I usually kill the cells I infect, whereas HTLV-I rarely does.

I: I can see how the ability to infect resting cells would be a great advantage.

HIV-1: It is advantageous. However, although I can infect some resting cells, very few virus particles are produced unless HIV-1–infected cells are proliferating. The reason is that for new viruses to be produced, integrated proviral genes must be transcribed into RNA by the cellular RNA polymerase. And when cells are resting, the goodies required for transcription are only produced at "maintenance" levels. What this means is that the proviral genome is "silent" in most infected cells—because the majority of helper T cells and macrophages are not proliferating. Of course, establishing reservoirs of latently infected cells is a critical feature of my plan to infect my hosts for as long as they live. So I shouldn't be too negative about latent infections.

VIRAL EVASION TACTICS

I: If your goal is to "infect for life," I imagine you invest heavily in tactics that allow you to evade host defenses.

HIV-1: Absolutely. I use a combination of "defensive" strategies, which allows me to evade the host's immune system, as well as "offensive" strategies, by which I actively subvert and then destroy the host defenses.

I: That sounds nasty!

HIV-1: It is nasty. Here's an example of what I mean by a defensive strategy. As I mentioned, I easily establish a stealth infection in which proviral DNA is integrated into the chromosomes of infected cells. Once this happens, the provirus can just sit there silently, biding its time. Because killer T cells do not recognize these cells as infected, the viral genetic information is safe from attack. In this respect, HTLV-I and HIV-1 infections are very similar: Both viruses establish a reservoir of infected cells which cannot be detected by the host immune system. This defensive posture is achieved very rapidly. It only takes five to ten days after the initial infection for me to reach a safe haven as an integrated provirus. In contrast, it takes one or two weeks for the adaptive immune system to be activated, and for killer T cells to be produced. Consequently, I usually can establish a latent infection before the antiviral weapons of the adaptive immune system can be deployed.

I: I would imagine that once you have a reservoir of undetectable, latently infected cells, the immune system must have a very difficult time dealing with an HIV-1 infection.

HIV-1: Yes, this feature almost always gives us the upper hand. In addition, HIV-1 can be passed efficiently from cell to cell without ever leaving the protected cellular environment—a process called transinfection. Neutralizing antibodies, which can prevent viruses from infecting their target cells, are only effective when viruses are traveling between cells, so these antibodies are totally useless when we transinfect neighboring cells.

I: So long as you are in a "defensive posture," as you call it, I can see that you would be pretty safe. But what happens when your latently infected cells reactivate and begin to produce virus?

HIV-1: It's true that the synthesis of new viruses eventually "blows the cover" for HIV-1–infected cells, and they do become targets for killing by T cells. However, I have a defense for that—rapid mutation. During each and every round of infection, my reverse transcriptase enzyme must produce a DNA copy of my RNA genome. This enzyme has the wonderful property that it makes about one error (mutation) each time it copies a piece of viral RNA. As a consequence, essentially all viruses produced in an HIV-1–infected cell are different from the virus which originally infected that cell. In fact, the body of an untreated person who is infected with HIV-1 generally contains a collection of HIV-1 genomes which include every possible single base mutation.

The reason this high mutation rate is extremely helpful to us is that HIV-1 can mutate so that a viral protein which formerly was recognized by killer T cells no longer can be recognized, or no longer can be presented by the MHC molecule those killer T cells were trained to focus on. When such mutations occur, that clone of killer T cells will be useless against cells infected with the mutant virus. Consequently, to destroy the mutant-virus–infected cells, a new clone of T cells which recognizes the mutant will have to be produced from scratch—and this will take a week or two. Meanwhile, the mutant virus, which has escaped surveillance by the obsolete killer T cells, is infecting more cells and making more virus. And every time it infects a new cell, it mutates again. The net result is that the HIV-1 mutation rate is so high that I can stay one step ahead of killer T cells or antibodies directed against me.

I: So three of the defensive tactics which make HIV-1 especially difficult for the host to deal with are: It can establish an undetectable, latent infection; it can spread stealthily from cell to cell; and it can mutate rapidly to "outrun" the adaptive immune system. That's very impressive. Any other tricks you use to antagonize host defenses?

HIV-1: Oh, I have lots more. I don't want to bore you with a list, but here's one of my favorite defenses. One type of cell I can infect is the plasmacytoid dendritic cell (pDC). I'm pretty sure humans came up with this cell type just to try to thwart retroviral infections—because these cells have pattern recognition receptors (TLR7) which can sound the alarm if the cell is infected by a retrovirus. Upon infection, TLR7 sends a signal to the nucleus of the pDC, and the genes for interferon α and interferon β are turned on, and interferon is produced. When IFN-α or IFN-β binds to interferon receptors on the surface of cells, it induces the expression of interferon stimulated genes (ISGs)—genes which have antiviral functions. One such gene encodes a protein, APOBEC3G, which is especially troublesome for retroviruses. This antiviral protein acts as a powerful mutagen: It introduces many mutations into the HIV-1 proviral genome during the process of reverse transcription. As I mentioned, the HIV-1 polymerase already is error-prone—and that's a good thing for us. However, the effect of the additional mutations introduced by APOBEC3G is, in many cases, to render the resulting mutant viruses non-functional. There is a limit to just how much any virus can mutate and still function.

I: I see what you mean about a host defense specifically designed to destroy retroviruses. That sounds serious.

HIV-1: It would be except that we have figured out a way to outsmart this defense. It's actually quite simple. To protect ourselves against "overmutation," the HIV-1 genome encodes an accessory protein, Vif, which targets that troublesome APOBEC3G protein for degradation in infected cells—neutralizing this host defense!

I: That is very clever, but isn't allowing the production of interferon a dangerous thing for you to do? After all, the interferon defense is quite powerful.

HIV-1: Powerful indeed. However, this brings us to the second way we defend ourselves—our "active" defenses, we call them. HIV-1 turns host defenses against the host, and uses those defenses to spread and maintain the viral infection.

I: That sounds insidious!

HIV-1: Oh yes it is. You mentioned the interferon system, so let me use that as an example. Probably the most difficult task we face is establishing the initial infection. Non-infectious virus particles outnumber infectious ones about 1000 to 1, and most HIV-1 infections are initiated when a single heroic virus manages to penetrate the mucosal barrier, avoid being eaten by macrophages which are patrolling the submucosa, and infect an immune system cell. Although HIV-1 can infect cells that are not proliferating, it doesn't do that very efficiently. Moreover, when resting cells are infected, they produce little or no new virus. After all, they are resting. In contrast, when HIV-1 targets helper T cells that are in the process of dividing, these cells can easily be turned into virus factories. The problem we face, then, is that most immune system cells are resting. It is only after they have been alerted that there has been an invasion that they become "activated" to play their respective roles in mounting a defense. What this means is that, in contrast to most other viruses, it is critical that HIV-1 alert the immune system to its presence—so that its target cells become activated.

I: That certainly is counterintuitive—that you would want to alert the immune system.

HIV-1: It <u>is</u> counterintuitive—heh, heh—and that's why it's so insidious. One of the ways we alert the immune system is by infecting those plasmacytoid dendritic cells. The interferon they produce not only warns cells of a viral infection, it also activates helper T cells and other immune defenders. And that's just what we need for efficient viral replication. So by infecting plasmacytoid dendritic cells, and by stimulating them to produce interferon, HIV-1 can activate resting immune system cells, making them targets for efficient infection and viral production.

I: I see what you mean about using the immune system to further your agenda. Any other examples?

HIV-1: Here's another one. Although the first cells we infect are in the submucosa, there are not a lot of immune system cells there. In fact, most of our target cells are found in lymph nodes and similar "secondary lymphoid organs." Consequently, a lymph node is about as close to heaven as a virus like me can get! The only problem is how to reach this "promised land." And here's the trick.

In addition to plasmacytoid dendritic cells, there is another type of dendritic cell which resides in the submucosa, the "conventional" dendritic cell (cDC). The job of this sentinel cell is to travel to a nearby lymph node to alert the immune system cells which reside there that there has been attack. That's how the system is set up to work—and it plays right into our hands. You see, HIV-1 virus particles can infect cDCs at the site of the initial infection, and we also can bind to the surface of these cells. Consequently, when cDCs head off to lymph nodes to report that there has been a viral infection, we just hitch a ride! In lymph nodes, there are lots of $CD4^+$ cells within easy reach, and because cDCs function to activate these cells, many of them will soon be proliferating—making them ideal candidates for HIV-1 infection. So by "commandeering" cDCs as our transportation vehicle, we turn the immune system against itself—and solve our problem of how to access lymph nodes. Said another way, by becoming cargo on traveling dendritic cells, HIV-1 takes advantage of the normal trafficking of immune system cells through lymph nodes, and turns the $CD4^+$ cells in these organs into virus factories.

I: That's sick!

HIV-1: Ah, but it gets even better! After I have used immune system cells for transport or for virus factories, I either disrupt their function, kill them outright, or make them targets for destruction by killer T cells. In fact, I destroy—one way or another—about 80% of the helper T cells in the lymph nodes which drain the intestine during the first three weeks of an HIV-1 infection! So I either directly or indirectly damage or destroy the helper T cells and dendritic cells required to activate the immune system and provide killer T cells and B cells with the help they need to function.

I: That really doesn't sound fair.

HIV-1: Fair? Don't you get it? Viruses don't care about "fair." We only care about four things: How can we access and enter the cells we want to infect? How can we reproduce in those cells? How can we evade host defenses long enough to reproduce? How can we spread within our host and be transmitted to other hosts? That's it! End of story. We are totally selfish.

HOW HIV-1 SPREADS

I: When I interviewed your cousin, HTLV-I, I learned that he has three ways of spreading within an infected human. I suppose you use these same three ways.

HIV-1: Yes, within our hosts, HIV-1 spreads by proliferation of infected cells harboring integrated proviral DNA, infection of cells by viruses produced by infected cells, and cell-to-cell transinfection. However, there are some differences. When HTLV-I–infected cells with integrated proviruses are activated, the usual outcome is a continued "stealth" posture in which the integrated proviral genome is quietly passed down to the resulting daughter cells, and little or no new virus is produced. In contrast, when cells latently infected with HIV-1 proliferate, the more common scenario is that viral RNA transcription is stimulated, and within about 12 hours, new virus particles begin to bud from the infected cell. These viruses then go on to infect other cells within the host. So the reservoirs of latently infected cells are very important to us, not only as a place to hide, but also as potential virus factories

I: And how do you spread from human to human?

HIV-1: HIV-1 virus and infected cells are found predominately in the blood, semen, vaginal secretions, and breastmilk of infected individuals—just as with HTLV-I. However, in contrast to an HTLV-I infection, in which these bodily fluids contain large numbers of infected cells, but very little infectious virus, the bodily fluids of an individual infected with HIV-1 contain <u>both</u> infected cells and infectious virus.

I: With HTLV-I, we saw that mother-to-child "vertical" transmission during breastfeeding was a major route for viral spread. Are you transmitted in the same way?

HIV-1: We can be, but it's a waste of time. The vast majority of children infected with HTLV-I at birth experience no disease symptoms throughout their lives, making it possible for them to transmit that virus to their young and also to their mates during a lifetime of sexual activity. This is not the case for us, since children infected with HIV-1 at birth almost never survive long enough to spread the virus by sexual contact—or by breastfeeding. So although about 25% of the children of HIV-1–infected mothers will be infected, mother-to-child transmission represents a dead end for us.

I: How about sex?

HIV-1: Sex is good, but not great. Although HIV-1 virions and HIV-1–infected cells are present in seminal fluid and vaginal secretions, sexual intercourse is not an inherently efficient way to spread HIV-1. In fact, the probability that an otherwise healthy man will infect a healthy female during a single episode of vaginal intercourse is only about one in a thousand. And the likelihood that a similarly infected woman will infect a healthy man is even smaller. What this means is that if HIV-1 were only transmitted either during childbirth or by men and women who had only one or a few sexual partners during a lifetime, HIV-1 could never become established in the human population. Transmission by these routes is just too improbable.

I: Aha! I'll bet that explains why HIV-1 has only become a health problem in the last few decades: Lifestyles have changed recently to include what we might term "urban practices."

HIV-1: Bingo! These "urban practices" have made it possible for HIV-1 to survive, and in some settings, to thrive. There are three principle core groups from which HIV-1 infections now spread: female prostitutes and the men who use them, men who engage in anal sex, and intravenous drug abusers. In Africa and in parts of Asia, most HIV-1 infections emanate from female prostitutes and their patrons. As large population centers emerged in Africa during the twentieth century, it became much more common for men to visit prostitutes. Moreover, many of these prostitutes are infected with the "usual" sexually transmitted diseases such as herpes, syphilis, gonorrhea, and *Chlamydia*. Some of these diseases cause ulceration of the tissues of the sex organs—a condition that makes transmission of HIV-1 much more likely. Others of these diseases cause inflammation of vaginal and penile tissues, resulting in the recruitment of many activated $CD4^+$ cells to these areas, increasing the number of local targets available for infection by HIV-1. So although by itself, HIV-1 is not spread efficiently during sex, in the context of other sexually transmitted diseases, the likelihood of sexual transmission of HIV-1 increases dramatically. The prevalence of the usual sexually transmitted diseases in the prostitute population set the stage for the efficient transmission of HIV-1 once prostitutes became infected with this virus. Customers of these prostitutes, equipped with both the usual sexually transmitted diseases and HIV-1, could then transmit these diseases efficiently to their wives and to other sex partners. It's a lovely story.

A second core transmission group is men who practice anal sex. Apparently, Mother Nature designed the anus to have things come out of it, rather than to have things put into it, because the mucosal tissues that line the rectum are prone to tearing during anal intercourse. Disruption of the integrity of the mucosal barrier increases the probability that HIV-1 can be transmitted, especially to the receptive partner. In addition, men who practice anal sex tend to have many sexual partners. Not only does this promiscuity increase the probability of spreading HIV-1 more widely, but the high multiplicity of sexual partners also increases the risk of contracting other sexually transmitted diseases—which can increase the chances of transmitting HIV-1. In the United States, it is estimated that among men, over 60% of HIV-1 infections were acquired during anal intercourse. Once infected, these men can spread HIV-1 infections not only to other men, but to women during intercourse. So men who practice anal sex have been very useful to us.

I: What about people who inject "recreational" drugs?

HIV-1: I was just getting to that. Intravenous drug abusers represent the third major focus from which we currently emanate. Many abusers share needles or syringes, and because persons infected with HIV-1 have large numbers of viruses and HIV-1–infected cells in their blood, this practice also results in the efficient "sharing" of HIV-1 infections. Indeed, about half of the women in the United States who are infected with HIV-1 contracted the virus during intravenous drug abuse.

I: What about people who are not in these "core" groups? How are they infected?

HIV-1: From these three foci of infection, HIV-1 is spread mainly by intimate contact. In addition to sexual dissemination, HIV-1 can be transmitted efficiently by contaminated blood used for transfusions or by contaminated blood products (e.g., clotting factor VIII). In fact, a person who receives a single unit of HIV-1–contaminated blood has greater than a 60% chance of becoming infected. Since 1985, blood and blood products have been screened for HIV-1 in the United States, so this route of spread doesn't work for us here anymore. However, in parts of the world where blood is not carefully screened, or where needles and syringes are reused, we are still in the "blood business."

So in contrast to HTLV-I, which thrives in small, isolated tribes, HIV-1 is a virus which prospers in an urban setting. Indeed, the "modern urban practices" of prostitution, anal sex, and intravenous drug abuse have allowed HIV-1 infections to "catch on." Not only that, but urban life provides a large number of uninfected hosts who live in close proximity with core transmission groups, making it more "convenient" for members of these groups to spread the virus to the general population. We love big cities!

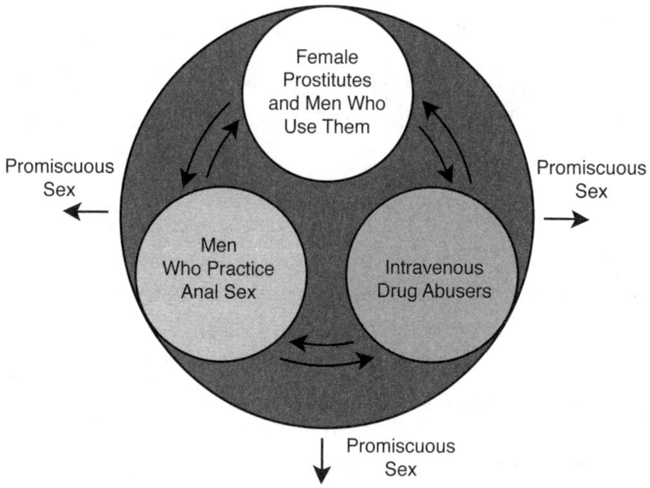

THE PATHOLOGICAL CONSEQUENCES OF AN HIV-1 INFECTION

I: So what happens to people who are infected with this "urban" virus?

HIV-1: First, there is usually about a 10 day "eclipse period" while we are reproducing out in the tissues, and hitching rides to nearby lymph nodes. Then there is a dramatic rise in the number of viruses in the body (the "viral load") as we multiply virtually unchecked by host defenses. As lymph nodes become centers of infection, they swell, and this lymphadenopathy is one of the most common early symptoms of an HIV-1 infection. In this acute phase of the infection our hosts frequently suffer from flu-like symptoms (fever, chills, muscle aches, etc.) that are caused by interferon and other cytokines produced as host defenses come on line.

Once the adaptive immune system reaches full strength, there is a marked decrease in the viral load as virus-specific killer T cells and antibodies go to work. This marks the beginning of the chronic phase of the infection.

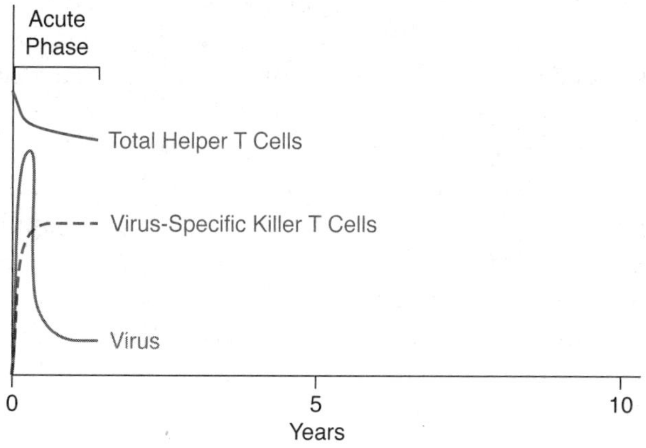

During the chronic phase, viral loads decrease relative to the levels at the height of the acute infection, but the blood of a chronically infected person usually still contains between 1,000 and 100,000 virus particles per mL. The number of virus-specific helper T cells and killer T cells also remain high—a sign that the immune system is still trying hard to defeat us.

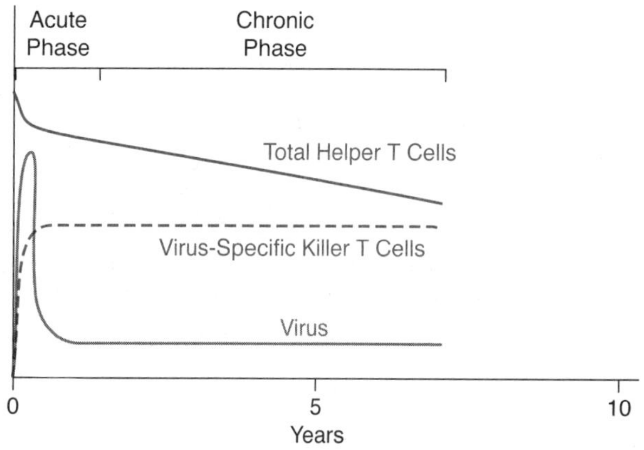

Throughout the chronic phase of an HIV-1 infection, there is a cycle of birth and destruction. Each day, over a billion new viruses are born, and many new $CD4^+$ T cells are produced, only to be infected by newly made viruses. Activated by the battle, these $CD4^+$ T cells synthesize more viruses, and are then destroyed either by the immune response or by the viral infection itself. There are about 100 billion $CD4^+$ cells which can be infected, and those that are "deleted" can be replaced by cells produced in the bone marrow. So these immune system cells really can be thought of as a "big organ."

As the chronic phase wears on, the total number of helper T cells slowly decreases as production of new cells cannot keep pace with the destruction of infected ones. In addition, we infect antigen presenting cells (e.g., dendritic cells), resulting in the death or malfunction of these cells. Because antigen presenting cells are essential for the adaptive immune system to activate both helper and killer T cells, putting these cells out of commission really helps our cause.

Eventually there are not enough helper T cells remaining to provide the assistance needed by virus-specific killer T cells. When this happens, the viral load increases dramatically, because there are too few killer T cells left to cope with newly infected cells.

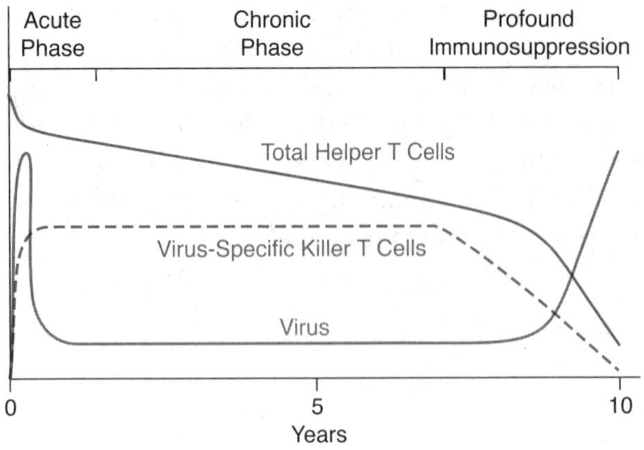

Eventually, the immune defenses are overwhelmed, leaving the patient open to unchecked infections by pathogens that would normally not be the slightest problem for a person with an intact immune system.

I: Such a sad story.

HIV-1: That depends on your point of view. During the chronic infection, huge amounts of virus are produced, but we are careful not to make our hosts too sick. Indeed, although a grand war is going on in the body of an infected human, an HIV-1 infection is relatively asymptomatic—at least in the beginning. This frequently gives us a decade-long "window" during which we can be transmitted to new hosts.

I: As I said, a very sad story. Is an HIV-1 infection uniformly fatal?

HIV-1: With many viruses, the acute phase of infection ends with "sterilization": All the invading viruses are destroyed, and memory B and T cells are produced which can protect against a later infection by the same virus. For a few individuals, an HIV-1 infection may also end either in sterilization or in long-term control of the infection. However, these "elite controllers" are the exception.

I: So you are a true killer.

HIV-1: Perhaps, but answer me one question: Suppose you humans were monogamous, didn't inject recreational drugs, and tested your blood supply for HIV-1. How long do you think I'd last?

I: Not very long.

HIV-1: Exactly. I'd be back to the chimps. So don't blame this on me. It's your "urban practices" which opened the door for HIV-1.

I: That may be true, but you certainly are different from your cousin, HTLV-I. In most cases, he seems to live peacefully with his human hosts. Why can't you be more like him?

HIV-1: Give me a break! Do you know how old he is?

I: I hear he has been around in the human population for about 100,000 years—although I'm not sure I believe that.

HIV-1: You can believe it. And how long have I been infecting humans? Less than 100 years. So he has had all that time to adapt—to work out a "standoff" with human hosts. I'm hoping I can do that too, but it will take some time. I'm new at this.

THE INTERVIEWER'S SUMMARY

HIV-1 infects immune system cells (e.g., helper T cells and dendritic cells) which have CD4 proteins on their surface. After binding to this protein and engaging a co-receptor molecule, the viral envelope fuses with the cell's plasma membrane, the single protein capsid opens, and two copies of the single-stranded RNA genome are released into the cytoplasm. This RNA is copied by the virus-encoded reverse transcriptase enzyme to yield a double-stranded, DNA provirus—which enters the nucleus of the infected cell. There, the viral integrase enzyme orchestrates the insertion of the provirus into one of the cell's chromosomes. This integrated provirus is then transcribed by the cellular polymerase (pol II) to create viral mRNA and new viral genomes. In the cytoplasm, two of these RNA genomes are brought together within a capsid constructed from virus-encoded proteins, and this complex then buds from the plasma membrane, picking up an envelope as it exits.

HIV-1 quickly establishes a reservoir of latently infected cells with integrated proviral genomes. In this latent state, the virus is very difficult for the immune system to detect. Latently infected cells can "reactivate" from time to time to produce new viruses—which can infect additional $CD4^+$ cells. The virus also can spread from infected to uninfected cells by cell-to-cell contact, thwarting the antibody defense. Importantly, the error-prone viral reverse transcriptase introduces mutations into the viral genome each time the virus infects a new cell, and these mutations usually can keep the virus one step ahead of the adaptive immune system. In addition to these defensive measures, the virus also goes on the offensive, commandeering the weapons of the immune system for its own ends, and disabling or overseeing the destruction of these weapons.

An HIV-1 infection can be transmitted during sexual intercourse, and the probability of transmission increases dramatically in the context of other sexually transmitted diseases. The virus also is efficiently spread to new hosts by the sharing of contaminated needles or by the transfusion of contaminated blood. The virus establishes a chronic, usually lifelong, infection during which the virus and the immune system continue to do battle. Unfortunately, the virus almost always wins.

GENERAL PRINCIPLES

1. Viruses which kill their hosts before they reach puberty cannot use vertical transmission to persist in the human population.
2. Retroviruses generally have difficulty infecting and reproducing in resting cells.

THOUGHT QUESTIONS

1. What is the "ancient" way in which HIV-1 was designed to spread (a trick question!)?
2. How is HIV-1 spread in "urban society"?
3. How does HIV-1 evade host defenses?
4. Discuss how HIV-1 can turn the immune system against itself by using processes that are essential for immune function to spread and maintain the viral infection.
5. Is HIV-1 well adapted to its human host?

Chapter 14

Herpes Simplex: A Virus That Hides

BACKGROUND

Although HIV-1's ability to establish a latent infection gives it a reservoir from which new viruses can emerge to sustain the infection, that virus also attacks the immune system head-on. In contrast, herpes simplex virus, after the initial infection, spends the majority of its life "in hiding," and none of its time actively attacking host defenses. Yet by maintaining a low profile, and taking advantage of humans' proclivity for intimate physical contact, herpes simplex virus has had great success. This virus currently infects about a third of the human population.

Two closely related herpes simplex viruses exist: HSV-1 and HSV-2. The genomes of these two viruses are about 50% identical, and it is believed that they diverged from a common ancestor about 9 million years ago. By the time they are old enough to vote, about 60% of all Americans will have been infected with HSV-1, whose preferred targets are the skin around the lips and the mouth. HSV-2 predominantly infects the genital area, and about 25% of the American population is infected with HSV-2. Although each virus has its favorite sites of infection, with the breakdown of taboos against oral sex, things have gotten rather mixed up. Now roughly 40% of all new cases of oral herpes are caused by HSV-2, while about the same fraction of new genital herpes cases involve HSV-1.

INFECTION BY HERPES SIMPLEX VIRUS

Interviewer: What cells do you target for infection?

Herpes simplex virus: During the initial infection, I target the epithelial cells that are located at the point of physical contact. I reproduce rapidly in those cells, and as they die, the viruses they produce are released into the surrounding tissues to infect more epithelial cells, amplifying the infection. These newly minted viruses also infect nearby sensory nerve cells.

I: Why would you want to infect nerve cells if you reproduce so efficiently in epithelial cells?

HSV: That's an important question. Although we don't reproduce efficiently in cells of the central nervous system, these infected neurons do provide a safe haven within which we are able to escape detection by host defenses.

I: I'll ask you about host defenses later, but for now, would you explain how you infect these cells?

HSV: Certainly. I'm an enveloped virus, and to initiate an infection, glycoproteins on my envelope (gB and gC) bind to heparan sulfate, a carbohydrate that is a "decoration" on the proteins that make up the outer surface of every cell. This binding isn't very strong, but it brings me close enough to my target so that another envelope glycoprotein (gD) can bind to receptor proteins (e.g., nectin-1) on the cell surface. Then, through the concerted action of several of my proteins, my viral envelope fuses with the cell's plasma membrane, and my DNA-containing viral capsid is released into the cytoplasm of the cell.

I: You mentioned that these heparan sulfate molecules are found on the surface of every cell. This must mean that you are not very selective about the cells you infect.

HSV: That's true. Although oral and genital herpes infections are the most common, we can infect any region of the skin or mucus membranes. In fact, we like to say that if humans enjoyed rubbing elbows as much as they like kissing and having sex, "elbow herpes" would be a common disease!

VIRAL REPRODUCTION

I: I suppose you get your name because you are a "simple" virus.

HSV: I beg your pardon! There is nothing simple about herpes simplex virus. I am a large, double-stranded DNA virus with enough genetic information to code for more than 80 proteins—proteins that are expressed in a carefully regulated temporal sequence. Some DNA viruses make do with only about a half dozen genes. They are the "simple" ones.

I: If other DNA viruses can get the job done so simply, why do you need all those extra genes?

HSV: The functions of many of the genes you call "extra" are required for three important features of a herpes simplex infection: the ability to reproduce in cells that are not proliferating, the ability to rapidly produce a burst of new virus in infected epithelial cells, and the ability to establish a latent infection in cells of the central nervous system. Even my structure is not simple. My linear DNA genome is enclosed in a protein capsid which is coated with an amorphous layer of proteins called the "**tegument**"—a structure which, I should point out, is unique to herpes viruses. Finally, the tegument-coated capsid is enclosed in an envelope supplied by the infected cell. And by the way, my tegument isn't just padding for the viral capsid. In fact, a number of the tegument proteins perform functions which are absolutely critical for the takeover of the infected cell.

I: So once you enter your target cell, what happens then? It sounds like you have a lot of "undressing" to do.

HSV: After I shed my envelope at the cell's plasma membrane, my partially disrobed virion makes its way to the cell nucleus, where replication of my genome will take place. There, my capsid "opens," and my DNA is injected into the nucleus, together with some of the tegument proteins.

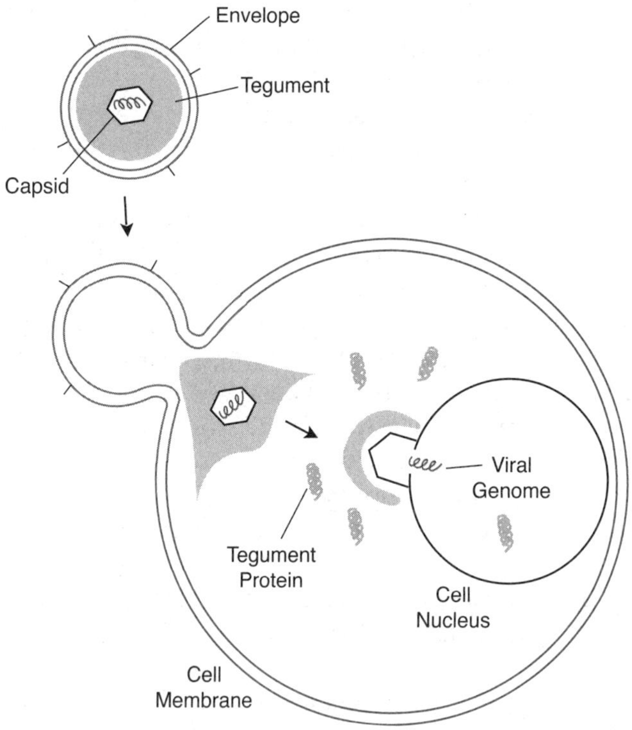

Inside the nucleus, my linear DNA molecule is circularized, early viral mRNAs are transcribed, and the DNA is replicated by a "rolling circle" strategy that produces many copies of the viral genome joined together on a long, linear DNA molecule. This giant piece of DNA is then used as a template for further viral mRNA synthesis before being cleaved to produce many copies of my linear viral genome. These newly made genomes are then encapsidated within the nucleus, and after tegument proteins are added, the encapsidated DNA (nucleocapsid) buds through the nuclear membrane acquiring a "temporary" envelope. As the nucleocapsid makes its way through the guts of the cell, this envelope is exchanged for another before the completed virus particle—composed of viral DNA, a protein capsid, tegument proteins, and an envelope—finally is released from the cell. Does that sound simple to you?

I: No, that's certainly not a simple process. You said that you can reproduce even in cells that are not proliferating. How do you manage that?

HSV: Many of the supplies required for DNA replication are present in limited amounts in human cells that are not proliferating. Since this includes most of the cells in a human, DNA viruses have evolved two different strategies to make it possible for them to reproduce in "resting" human cells. Some DNA viruses (e.g., the adenovirus) encode proteins that force resting cells to proliferate, thereby insuring that the enzymes and building blocks required for efficient viral reproduction will be available for their use. However, because the coding capacity of my genome is so great, I am able to employ a different strategy. I actually have genes which encode the proteins for DNA replication that are limiting in resting cells. For example, in addition to encoding my own DNA polymerase, I also encode my own version of thymidine kinase, an enzyme required to produce the large quantities of the building blocks needed for the synthesis of new viral genomes.

I: Does all that preparation take a long time?

HSV: Not at all. I actually reproduce quite rapidly in epithelial cells. One reason I am so fast is that I encode a protein, ICP27, which inhibits splicing of cellular mRNAs. Because most herpes viral mRNAs lack introns, and therefore don't need to be spliced, inhibition of mRNA splicing helps focus cellular protein synthesis on the production of viral proteins. In fact, within the "hijacked" cell, viral reproduction is so rapid that in about 24 hours, a single infected cell can produce thousands of new viruses. This massive amount of viral reproduction isn't so great for the herpes-infected epithelial cell, however, and it eventually dies from the exertion.

I: Do you also kill the nerve cells you infect?

HSV: Certainly not. Although I turn epithelial cells into doomed virus factories, I gently infect nerve cells. My goal is to establish a stealth, latent infection in cells of the central nervous system—an infection that will last a lifetime. Then, from time to time, the viruses in these latently infected cells can "reactivate" to facilitate my spread to new hosts.

EVASION OF HOST DEFENSES

I: Since your aim is to establish a lifelong infection, you must have figured out ways to defend yourself against host antiviral defenses.

HSV: Yes, indeed I have. In fact, about 50% of my coding capacity is devoted to this important task. I know you don't want to hear about them all, so I'll just mention a couple of evasion tactics with which you may not be familiar. If you want the complete list, you can always buy one of those big, thick, virology textbooks—if you can afford one!

I: They <u>are</u> expensive.

HSV: My genome is rich in unmethylated CpG dinucleotides, which are not abundant in cellular DNA. This "abnormal" DNA can be detected by TLR9 when I infect plasmacytoid dendritic cells that are guarding the epithelial surfaces. These Toll-like receptors then initiate a kinase cascade which results in the transcription of the genes for IFN-α and IFN-β.

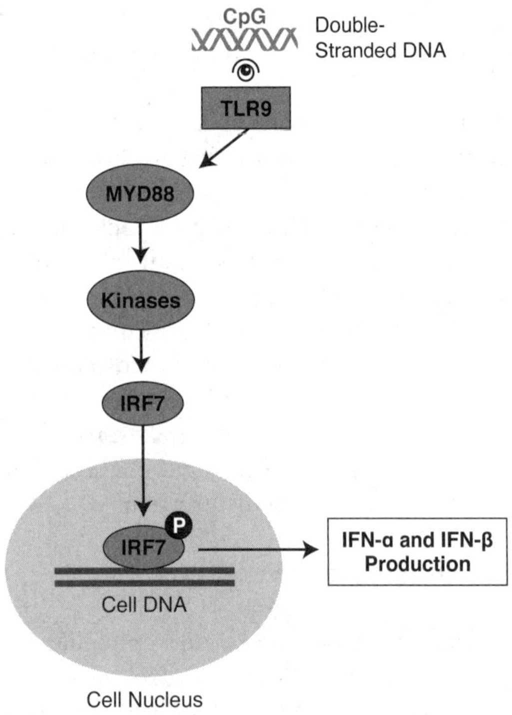

The interferon produced by these plasmacytoid dendritic cells can bind to interferon receptors on nearby cells and activate the expression of interferon stimulated genes. Consequently, I have had to evolve ways of protecting myself from the effects of these dangerous, antiviral genes. For example, one of the interferon stimulated genes, PKR, can bind to double-stranded RNA and phosphorylate eIF2α, a protein required to initiate protein synthesis. Unfortunately, when eIF2α is phosphorylated, protein synthesis ceases, leading to the death of the infected cell—and the viruses within it. Although I am a double-stranded DNA virus, I use overlapping regions of both strands to synthesize some of my viral mRNAs—and the resulting double-stranded RNA would activate PKR. That would not be good. So to protect myself from this fate, I encode a "defensive" protein, ICP34.5. This protein is synthesized early in an infection, and it orchestrates the dephosphorylation of eIF2α, allowing protein synthesis to continue.

Another danger I must protect myself against is an attack by the complement system. When we viruses are traveling between our target cells, complement proteins can attach themselves to our surface. Having those proteins hanging on us is like waving a red cape in front of a bull—because they function as "eat me" signals which stimulate ingestion and destruction by macrophages. Fortunately, I'm ready for this. One of the viral proteins that makes up my envelope (gC) binds to the complement proteins and disrupts their function.

I: And what about the adaptive immune system? How do you counter that defense?

HSV: Yes, that's a problem too. Once the adaptive immune system has been activated, virus-specific antibodies are produced. Normally these antibodies would tag (opsonize) us for phagocytic ingestion, providing "backup" for complement proteins in that role. However, my envelope includes proteins that bind to the Fc region of IgG antibodies—the part of the antibody which normally would attach to a macrophage. As a result, the antibody no longer can form a "bridge" between me and that terrible phagocyte to help him eat me. Because phagocytosis of **opsonized** virus is one of the major ways the immune system deals with viruses which are outside of cells, interfering with opsonization by both complement proteins and IgG antibodies is a very effective way to evade this defense.

I also have ways of avoiding destruction by killer T cells. For example, within three hours after I infect a cell, I oversee the production of a viral protein, ICP47, which interferes with the TAP transporter system, causes the degradation of class I MHC molecules, and makes it less likely that killer T cells will recognize infected cells and destroy them (and me!).

I: With all these defensive tactics, I would think that you would be "untouchable" by host defenses.

HSV: I wish! I invest a huge amount of "genetic energy" to defend myself, and yet, within about two weeks, all the infected epithelial cells are killed, either by the virus infection, or by the immune response to it. It's very frustrating to work that hard and achieve so little.

I: But clearly that can't be the entire story. After all, how successful would a virus be which could only infect humans, was mainly spread by intimate contact, and could only be transmitted during the first two weeks after infection? Some wannabe viruses may have tried that strategy, but they didn't survive to tell the tale. So what's your answer to this conundrum?

HSV: In a word, I hide.

I: That's two words, but I get the point. Is that where the nerve cells come in?

HSV: How perceptive. Yes, to avoid destruction, I establish a latent infection in nerve cells. During the latent infection, few, if any, herpes-encoded proteins are produced. This is in striking contrast to the situation during the **productive infection** of epithelial cells—in which more than 80 viral proteins are made. The killer T cell, which recognizes viral proteins displayed on infected cells, is the main weapon used by the immune system to destroy infected epithelial cells. However, these killers can't detect a protein that isn't made. So the paucity of viral protein synthesis in nerve cells makes it difficult for killer T cells to recognize virus-infected neurons. Moreover, nerve cells normally produce relatively few class I MHC molecules, so the small number of viral proteins that are made must compete with the much larger number of normal cellular proteins for presentation on a limited number of MHC molecules. The end result is that latently infected neurons are rarely destroyed by killer T cells. Consequently, latently infected nerve cells give me a safe place to hide.

I: So herpes simplex virus requires two different kinds of host cells for its lifecycle: epithelial cells for rapid viral reproduction, and nerve cells in which to hide by establishing a latent infection. That's in contrast to HIV-1, which can either productively or latently infect the same kind of cell—a $CD4^+$ T cell.

HSV: I suppose. But I'm not an expert on the HIV-1 lifestyle.

HERPES SIMPLEX VIRUS' STRATEGY FOR VIRAL SPREAD

I: I'm not either, but I just interviewed that virus, so I remember a few things. For example, when HIV-1 establishes a latent infection in $CD4^+$ cells, the genetic information of the virus is integrated into the chromosomes of the infected cells. As a result, when these cells proliferate, the integrated proviral DNA is passed down to daughter cells, contributing to viral spread within the infected individual. I presume you use this same strategy when you latently infect nerve cells.

HSV: No, not at all. During latency in neurons, my viral genome exists as a "free floating" piece of DNA which is not associated with a host chromosome. This makes perfect sense. Once sensory nerve cells have taken their places in the human body, they no longer proliferate—so if my DNA were integrated into the genome of a nerve cell, I'd be trapped there with no way to spread either within that host or to new "clients." No, integration of my viral DNA into a cell's chromosomes would not be smart.

I: I see your point. But then how do you manage to spread?

HSV: I go through cycles of hiding in nerve cells and reproducing in epithelial cells. It's so peaceful and quiet in those nerve cells. I really wish I could just stay there. However, I must spread to other hosts to survive. Consequently, I'm compelled to abandon my latent state from time to time and "reactivate." Reactivation is controlled in part by viral genes that are expressed within latently infected nerve cells. In addition, episodes of reactivation usually are triggered by "external factors": psychological stress, physical trauma (e.g., friction during sexual intercourse), fever, ultraviolet light, etc. When reactivation occurs, the virus that is hiding in the nerve cell "wakes up," reproduces to a limited extent, exits the neuron near where it entered, and infects epithelial cells to produce many new viruses.

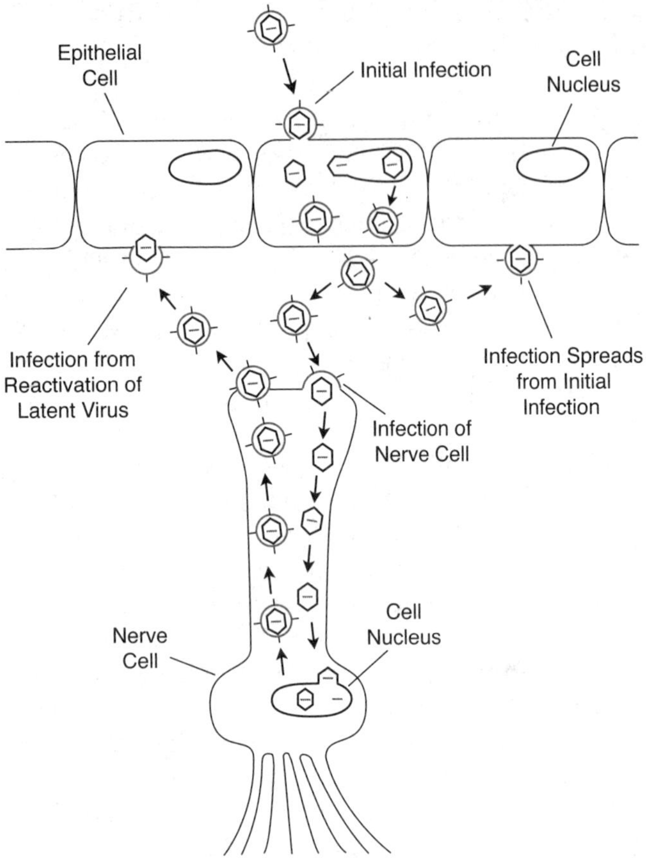

I should also point out that I don't infect just any old epithelial cell when I reactivate. I only infect epithelial cells which are in close proximity to the latently infected nerve cells. As a result, herpes simplex infections usually are very localized. This provincialism is in my best interest: A systemic herpes simplex infection would almost certainly kill my host before I could spread to another human.

I: So how do you spread?

HSV: Herpes simplex virus is transmitted when virus produced in the epithelial cells of an infected individual comes in contact with the epithelial cells of a recipient. Obviously, this can occur by "skin-to-skin" contact. However, I also can be transmitted when virus-containing saliva or genital secretions contact either epithelial cells of the genital/oral mucosa or epithelial cells in abraded skin—skin in which the protective layers of karatinized cells are broken to allow me access to the underlying epithelial cells. Indeed, this mode of transmission is quite efficient: A woman having sex with a man with active herpes has an 80% chance of contracting the disease during a single encounter. Moreover, by establishing a latent infection in neurons from which I can be reactivated from time to time, I am able to extend my infectious period over the lifetime of my human host. My strategy of infection, latency, and reactivation clearly works very well: Over 80% of all Americans born today will be infected with HSV-1 or HSV-2 or both. I'm a very popular virus.

PATHOGENESIS

I: So tell me this: With all this hiding and infecting, what sort of damage do you do to your hosts?

HSV: Okay. You're a smart fellow. On the basis of what I have told you about my lifestyle, what would you predict?

I: Ah, a common trick—turning the question back on the interviewer! But I'll give it a shot. You've told me that herpes simplex virus reproduces rapidly in epithelial cells of the skin or mucosal surfaces, destroying these cells, and producing a large burst of virus that can infect neighboring epithelial cells. From these considerations, I would predict that one pathological consequence of a herpes simplex infection would be the ulceration of infected epithelial cell layers caused when these cells either are killed by the virus itself or by the immune response to the infection. Consequently, I would think that for most individuals, a herpes simplex infection would result in an annoying (and painful), recurring skin disease.

HSV: A well thought-out answer. I'll bet you've had some personal experience with this "annoying skin disease," as you call it.

I: Yes, I have.

HSV: But I'll bet you didn't know this: Most humans seem to think that herpes simplex virus only is "shed" by infected individuals during reactivation events that lead to ulcers or blisters. Not so. Many reactivation events are asymptomatic—being rather quickly dealt with by the immune system. This means that virus shedding is more or less a continuous process. So in a practical sense, a herpes simplex infection probably should be regarded more as a chronic infection than as a latent infection which only reactivates periodically. What I like about this feature is that, because the infection can be transmitted by asymptomatic individuals, the probability of viral transmission is increased substantially—because persons with blisters or sores are likely to avoid intimate physical contact.

I: I can see why you might like that. But besides the blisters, are there other pathological consequences of a herpes simplex infection?

HSV: There are a couple. A herpes infection can be spread by the fingers of an infected individual from lesions around the lips to the epithelial cells of the cornea of the eye. Indeed, about 300,000 new cases of ocular herpes are diagnosed each year in this country, and these infections sometimes lead to blindness.

I: So people with a herpes simplex infection should be careful not to rub their eyes!

HSV: Yes, they should. But as I'm sure you understand, infecting someone's eyes is not something we try to do. It is an unintended consequence. "Seeing eye-to-eye" is a euphemism. Spreading that way is very unlikely.

I: Any other pathological consequences that we should know about?

HSV: There is one more. But again, I'll give you a chance to figure it out. What do you think would happen if a person who is immunosuppressed becomes infected with herpes simplex virus?

I: Here we go again! Okay, I'd predict that, because it is the immune response which limits most herpes infections, a person with a compromised immune system would experience more frequent reactivation episodes with more severe consequences.

HSV: That's true. For example, newborns, because of the immaturity of their immune system, are especially susceptible to devastating herpes infections. These infections most frequently occur during birth when the baby is bathed in virus-containing genital secretions from an infected mother. The risk of infection is greatest (about 30%) if the mother is experiencing a primary infection during delivery. In addition to the usual sites of infection (mouth, eyes, and skin), the virus can spread, relatively unchecked by the immune system, to the central nervous system and to organs such as the adrenal glands, lungs, and liver. Unfortunately, neonatal herpes is frequently lethal. But again, it is an unintended consequence of the baby's mother being infected. We certainly would not infect babies on purpose. After all, that baby isn't going to be kissing or having sex any time soon!

I: No, I don't think he will be.

THE INTERVIEWER'S SUMMARY

Herpes simplex virus relies on a "two-cell" strategy to persist in the human population. Viruses produced in the epithelial cells of an infected person are spread to a new host by intimate physical contact. After a brief period of reproduction in the new victim's epithelial cells, the virus then infects nearby neurons, and establishes a latent infection—a "silent" infection from which the virus can reactivate from time to time to infect more epithelial cells, and spread to new hosts. This rather complicated lifestyle is possible because herpes simplex is a very large, linear, double-stranded DNA virus with genes for more than 80 proteins. This genetic richness even allows herpes simplex virus to reproduce in resting cells.

The viral genome is protected by a protein capsid, surrounded by a layer of tegument proteins, and enclosed in a cell-derived envelope. When the envelope fuses with a target cell's plasma membrane, the tegument-coated capsid is released into the cytoplasm. The DNA genome is then "injected" into the nucleus of the cell, where the virus uses a "rolling circle" process to replicate its genome—a method of DNA replication which bears little resemblance to that used by the cell to replicate its chromosomal DNA.

Herpes simplex virus devotes about half of its genetic information to evading host defenses. In addition to establishing a latent, virtually undetectable infection of nerve cells, the virus encodes proteins which interfere with the opsonization of virions by complement proteins and antibodies. Viral proteins also can protect against the effects of the interferon system, and can compromise the ability of class I MHC molecules to present viral proteins to killer T cells.

Despite all these countermeasures, the immune system usually takes less than two weeks to destroy all the infected epithelial cells. The strong immune response forces the virus to "retreat" to the safety of nearby nerve cells, and the destruction of the epithelial cells results in the blisters that are characteristic of a herpes simplex infection. At a later time, after immune defenses have "relaxed," the virus reproduces to a limited extent in the nerve cells, infects another "set" of nearby epithelial cells—and the saga continues for the life of the infected individual. Because it is the immune system which "controls" the infection, the consequences of a herpes simplex infection for a person with an immature or compromised immune system can be life-threatening.

The virus usually is spread from host to host by vaginal or oral sex. However, transmission can take place whenever infected epithelial cells of one person are brought near enough to the epithelial cells of another person to allow the transfer of virus. In fact, one of the pathological consequences of a herpes simplex infection is ocular herpes, which can result when a person touches a blister on his lip and then rubs his eye.

GENERAL PRINCIPLES

1. Viruses have two ways they can deal with the difficult problem of infecting resting cells. Some viruses solve this problem by producing proteins which give the cell a "kick" to make it proliferate. Other viruses provide those components of the replication machinery which are missing, or are in short supply, in a resting cell.
2. Most viruses (even ones with double-stranded DNA genomes) generate some double-stranded RNA when they reproduce – RNA which can alert cellular pattern recognition receptors to the presence of the virus.

THOUGHT QUESTIONS

1. We have examined one other DNA virus, adenovirus, which is able to infect resting cells. Compare the strategies that herpes simplex virus and adenovirus use to solve the "resting cell problem."
2. HTLV-I and HSV both are able to establish latent infections. Compare the strategies these viruses use to accomplish this feat.
3. We have discussed three viruses – measles, hepatitis A, and herpes simplex – each of which requires multiple cell types to carry out its reproductive program. Compare the lifestyles of these three viruses.

Chapter 15

Human Papillomavirus: A Very Quiet Virus

BACKGROUND

The final participant in our Parade of Viruses is the human papillomavirus (HPV). As defined by differences in their major capsid proteins, there are about a hundred different genotypes of human papillomavirus. Each genotype infects specific regions of the human body, and some are spread by activities as casual as shaking hands or walking on the deck of a swimming pool. Other genotypes usually are spread by intimate physical contact, and they will be our focus here.

In the United States, the incidence of genital HPV infections has increased dramatically over the last three decades, rising at a rate faster than that of genital herpes infections. In fact, human papillomavirus is now second only to *Chlamydia* as the most commonly acquired sexually transmitted disease, with at least 20 million Americans currently infected.

Infection with certain oncogenic types of HPV dramatically increases the risk of cervical cancer. Worldwide, roughly 500,000 new cases of cervical cancer are diagnosed each year, and in the United States, about 5,000 women die of this disease annually. Despite its obvious medical importance, the details of how this virus reproduces are sketchy, because it has been very difficult to observe viral reproduction in the laboratory. In fact, much of what is thought to be true about HPV reproduction has been extrapolated from the habits of closely related DNA viruses which do reproduce well in cultured cells (e.g., the monkey virus, SV40).

INFECTION

Interviewer: Which cells do you infect and how do you access them?

Human papillomavirus: We human papillomaviruses are extremely picky about the cells we infect and the conditions under which we reproduce.

I: Could you please talk a little louder? I can hardly hear you.

HPV: Sorry. I'm a very quiet virus—but I'll try to speak up. As I was saying, we human papillomaviruses are extremely picky about the cells we infect and the conditions under which we reproduce. Our targets for infection are the "basal cells" which are located beneath all skin and mucosal surfaces. These cells are attached to the basement membrane, and include basal stem cells which proliferate on demand to replace epithelial cells as they are lost or damaged. These stem cells proliferate slowly, and when they divide, some of the progeny remain attached to the basement membrane and become stem cells themselves, whereas others are pushed upward toward the surface. Once they disengage from the basement membrane, the stem cells mature (differentiate) and stop proliferating. That's important for my lifestyle.

Driven toward the surface by the continued basal cell proliferation below, the maturing epithelial cells of the skin (the keratinocytes) dedicate themselves to the production of keratin proteins—much as maturing red blood cells become "factories" which produce hemoglobin. As the keratinocytes approach the skin surface, they flatten and eventually die, exhausted from the all-out effort of producing huge amounts of keratin proteins. When they die, their nuclei break down, and the cells become flattened bags of keratin. These dead cells, usually 10 to 20 layers deep, function as interlocking "shingles" which provide protection against the outside environment. When, as the result of wear and tear, these dry shingles flake off the skin surface, they are replaced by a new set of shingles rising from below—the end result of basal cell proliferation. Usually it takes several weeks for the daughters of basal stem cells to reach the skin surface and become household dust.

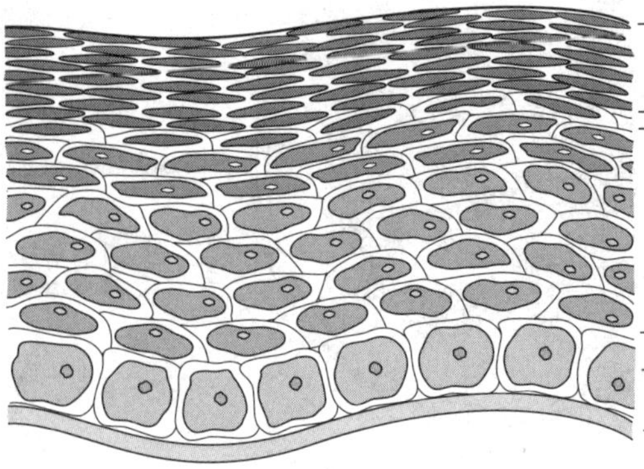

For the mucosal surface that lines the vagina, the story is a bit different. As the daughter cells are pushed higher and higher by proliferating basal cells, they flatten out, but they remain alive with their nuclei intact. In addition, the epithelial cells of mucosal surfaces do not produce huge amounts of keratin proteins, so the cells at the top of the stack (the squamous cells) are more like moist pancakes than dry shingles. As the squamous cells in the upper layer slough off, they are replaced by others rising from below.

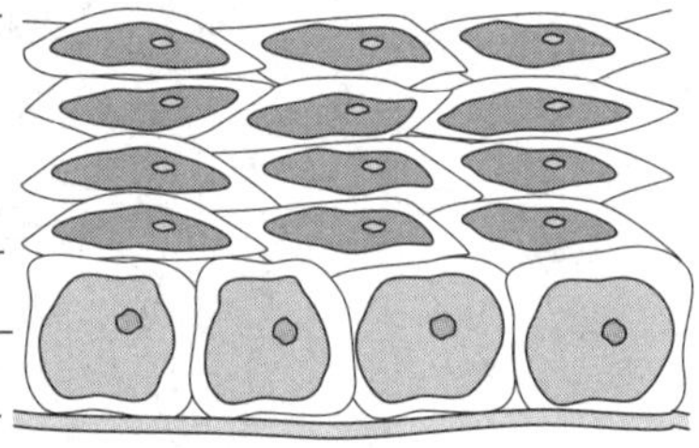

I: That's all very interesting, but I asked you how you access and enter your targets.

HPV: I understand your question, but to appreciate the amazing lifestyle of a human papillomavirus, you need a little background. To continue, during an infection, we human papillomaviruses make our way through cracks in the skin or tears in the mucosal barrier until we reach our target—the basal stem cell. One reason we choose these cells is that they are proliferating. The human papillomavirus is a double-stranded DNA virus with only eight genes, so there is no way we can provide all the materials required to replicate our genome. Consequently, we must rely heavily on target cells which are proliferating and which, therefore, can provide most everything we need to reproduce. Even so, when we infect basal stem cells, the infection only produces a small number of viral genomes (generally fewer than 50 per cell). Moreover, we do not complete our reproductive program in these basal cells, and no capsid proteins or virus particles are produced.

I: But doesn't that leave you trapped inside the basal epithelial cell?

HPV: So it would seem, but there's more to the story. As the infected cells move upward toward the surface and begin to mature, they normally would stop proliferating. However, I encode two viral proteins, E6 and E7, which keep the maturing epithelial cells in a "proliferative mode," so that cellular genes required for DNA replication continue to be expressed, and the "brakes," which normally would cause these cells to stop proliferating, are released. In these infected cells, the rate of viral DNA replication increases, many more viral genomes are produced, viral capsid proteins are made, and new virus particles are assembled. By the time the virus-infected cells reach the "top floor," many new virions will have been produced, and these are shed with the skin, or from the mucosal surface. So here's the strategy: By "shifting gears" as epithelial cells mature, the human papillomavirus goes from simply maintaining its genome in infected basal cells to a full-blown, productive infection in more differentiated epithelial cells.

I: So you solve the "resting cell problem" by encoding proteins that keep the maturing epithelial cells from shutting down their replication machinery. Very clever.

HPV: Yes, but it's not as easy as it sounds. My E7 protein binds to a cellular protein, pRB, whose usual job is to put the brakes on proliferation. The binding of E7 to pRb releases this proliferation block, and leads to the activation of a whole set of cellular genes whose expression is required for DNA replication. But this creates a potential problem. Although E7 does function to make the infected cell "replication ready," the unscheduled release of the pRB-enforced replication block is viewed by the cell as something abnormal. And when abnormal cellular DNA replication takes place, cells are programmed to commit suicide by apoptosis.

I: Yes, I've heard that the execution of this suicide program is one of the host's best defenses against viruses which tinker with the cell's replication machinery.

HPV: That's true, because when the cell commits suicide, the viruses inside die with it. Fortunately, I've solved that problem. To keep the cells I infect from dying, I encode another protein, E6, which thwarts the suicide defense. So by working together, my E6 and E7 proteins can trigger unscheduled cellular DNA replication—so I can reproduce—yet avoid the usual consequence of disturbing the cell cycle: death by apoptosis.

I: Well done! But you didn't mention how you enter those basal epithelial cells.

HPV: That's something I have to be quiet about. I don't want to tip off the virologists—who really aren't sure how I enter my target cells. Currently, they think that my major capsid protein, L1, binds to heparan sulfate molecules on the cell surface, and that this attachment changes the conformation of my capsid—which then allows my capsid proteins to bind to an unidentified entry receptor, resulting in viral endocytosis. Maybe yes, maybe no. I'll just leave it at that.

VIRAL REPRODUCTION

I: So after your still-mysterious entry into your target cells, how do you reproduce?

HPV: I'm a small, circular, double-stranded DNA virus with a single capsid made of protein. After entry, my protective capsid is partially removed, and my viral DNA, still cloaked in some of my capsid proteins (e.g., L2), enters the nucleus of the cell where replication of my genome takes place. Six of my eight genes are expressed early after infection, and they encode proteins that are involved in regulating viral mRNA synthesis and in replicating my genome. My other genes encode the two proteins (L1 and L2) used to construct new viral capsids.

I take advantage of the cell's polymerase to synthesize my mRNAs, and all of my messenger RNAs are transcribed from only one strand of my double-stranded DNA genome. It might appear to you that I have made a mistake here. After all, I could easily have expanded the coding capacity of my tiny genome by using both DNA strands to code for proteins—a trick which some other viruses use to advantage. However, avoiding the transcription of overlapping genes is critical for establishing my quiet lifestyle.

After my DNA is replicated, each new viral genome is enclosed in a protein capsid, and the new virus particles leave the nucleus and exit the cell. There is another point about my reproductive style I'd like to stress. When I infect my target cells and my viral DNA reaches the cell's nucleus, my genomes remain "episomal": They float free in the nucleus of the infected cell, unassociated with host chromosomes.

VIRAL EVASION TACTICS

I: You mentioned that you lead a "quiet life." Is that to protect yourself from host defenses?

HPV: Yes, that's exactly why I maintain a low profile. Because all of my mRNAs are transcribed from the same strand of viral DNA, I don't synthesize telltale, double-stranded RNA when I reproduce. Consequently, a human papillomavirus infection results in the production of very little interferon. Also, I infect basal epithelial cells very gently. They only produce a small number of viral proteins, and I'm careful not to kill those cells—so that the immune system is not alerted by unusual cell death. In fact, because I induce very little interferon production, and do not kill my target cells, the immune system usually is clueless that there has been an HPV infection.

I: I can see how you might lay low in the basal epithelial cells, but when you begin to produce virions on the way up to the surface, there must be a huge number of viral proteins produced—for example, to construct all those capsids. Doesn't that tip off the immune system?

HPV: Bravo! You've just discovered the genius of my lifestyle! In dealing with the immune system, location is everything. At the basal level, there are lots of immune system cells just waiting to detect a viral infection. So when the cells I infect are down there, I must be very quiet indeed. However, as the infected cell travels upward, I leave all those nasty immune warriors behind. Consequently, when I switch from latent mode to reproductive mode, I can be just as loud as I wish. There simply aren't any immune system cells there to "hear" what's going on.

I: But doesn't the immune system ever become aware of your infection?

HPV: Yes, I probably overstated my case a bit because most HPV infections are cleared by the immune system within a year or two. Even though I am very quiet, the innate and adaptive systems usually do eventually "wake up." What's most annoying is that this wake-up call usually has nothing to do with me. Other sexually transmitted diseases (e.g., *Chlamydia*) sometimes cause an inflammatory reaction in the nearby epithelium, activating both an anti-*Chlamydia* and an anti-HPV immune response.

And then there are the warts. I guess I'm most famous as "the wart virus," although that's not how I'd like to be remembered. You see, causing warts really is not in my best interest. I'm not spread any more efficiently from the surface of a wart than from any other region of an infected epithelium. And if warts are abraded, the resulting inflammation (usually caused by superinfecting bacteria) may trigger the immune response—and blow my cover. Yes, I certainly could do without the warts.

VIRAL SPREAD

I: I could too. But tell me about how you spread from host to host.

HPV: Here I have to brag a little. Basal stem cells are self-renewing, so infecting them allows me to establish a long-term reservoir of infection. Then, by producing infectious virus in epithelial cells during my "elevator ride" to the surface, a continuous supply of newly made virus is delivered to the "exterior" where it can be transmitted to other humans. Human papillomavirus spreads to new hosts either by direct physical contact, or when a person rubs up against a surface on which the virus has been deposited.

I: That's a clever strategy. I suppose you also can spread from an infected mother to her child. That certainly qualifies as "direct physical contact."

HPV: You're right. An HPV infection can be spread vertically, but that route of transmission is rather inefficient in comparison to sexual intercourse—my favorite way to spread. I'm not like those noisy respiratory viruses which love crowds and coughs and sneezes. I prefer quieter, more intimate settings—like the back seat of a car.

I: I see.

HPV: I'd also like to point out that the spread of human papillomavirus infections is facilitated by the fact that my viral capsid is impervious to agents which would destroy many other viruses: detergents, acids, ether, heat, and drying. In fact, the detergents used in condoms, which can damage or destroy HIV-1 and herpes simplex viruses, are helpless against me. I may be quiet, but I'm tough!

I: You certainly do sound like a tough virus, but you mentioned earlier that you usually don't infect your hosts for life—so you must have a weakness.

HPV: Yes, I do. In contrast to herpes simplex infections, which generally are lifelong, most genital human papillomavirus infections are transient, persisting for a few months to a few years. There are two main reasons for this. First, when basal epithelial cells proliferate, their cellular DNA is carefully apportioned, so that each daughter cell gets a complete set of chromosomes. In contrast, episomal HPV DNA is meted out to daughter cells on a random basis as the infected cells proliferate. Consequently, a daughter cell that is destined to mature and "ride the epithelial elevator" may receive all the viral genomes, while the daughter cell which is to become a "reservoir" stem cell, may be left uninfected. So one reason why I usually don't cause lifelong infections is that, over time, my genome can be "lost" from infected basal stem cells as they proliferate. The second reason for the transient nature of most HPV infections is that once we are detected, we usually are rather easily eradicated by the immune system. When found out, we go quietly.

I: But you currently infect millions of Americans, and are the second most "popular" sexually transmitted virus. If you have such a short infectious period, how do you infect so many?

HPV: Well, first of all, you humans love having sex. So a year or two really isn't that short a time in which to spread from one person to another. But there's also a trick we human papillomaviruses play. I'll tell you our secret.

It is true that when the immune system finally is activated, our visit with that host is over. However, this immune destruction by killer T cells is HPV genotype specific. Consequently, when killer T cells destroy cells infected with one genital HPV genotype, other genotypes may be left untouched. Because there are more than a dozen different genotypes which can infect the genital areas, people who are infected with one of these genotypes frequently are also infected, or will subsequently be infected, with other HPV genotypes. So by acting like a "tag team" in a wrestling match, viruses that "individually" establish transient infections can "collectively" end up infecting a large fraction of the population at any given moment.

HPV-ASSOCIATED PATHOLOGY

I: Since you are known as "the wart virus," I suppose that warts are the most frequent outcome of a human papillomavirus infection.

HPV: Actually not. In contrast to herpes simplex virus, which kills the epithelial cells it infects and causes blisters, human papillomavirus does not kill its target cells, and most HPV infections are unapparent. Although warts are the most visible manifestation of a genital HPV infection, genital warts are a relatively uncommon outcome. Indeed, genital warts are usually just the tip of the iceberg. For example, when a large number of college-age women were tested, almost half were found to be infected with genital HPV, yet only about 2% of the infected women had genital warts.

I: If the worst you do is cause a few genital warts, I'm not sure why I'm interviewing you. There must be something you haven't told us.

HPV: The HPV genotypes (HPV-6 and HPV-11), which normally cause genital warts, can be "transplanted" to the respiratory tract during oral sex. In addition, although vertical transmission is not a very efficient way for me to spread, the respiratory tract of an infant can be infected if that child is born to an HPV-infected mother. Respiratory HPV infections can progress from the vocal cords downward toward the lungs, and the resulting warts can block the airway if untreated.

I: Any other pathological consequences of a human papillomavirus infection that you haven't told us about?

HPV: Hum? Let me see... Well, there is the cancer.

I: Cancer?

HPV: Yes, in rare cases, infection with certain "oncogenic" types of human papillomavirus can result in cervical cancer (mainly cervical carcinoma). However, less than 1% of the women who are infected with genital HPV will ever suffer from cervical cancer. It's really rare, and causing cervical cancer is certainly not something I do on purpose—otherwise it wouldn't be so rare.

I: Less than 1%, you say? But aren't millions of women currently infected with human papillomavirus?

HPV: Yes, they are. In fact, although only a tiny fraction of infected women ever get cervical cancer, so many women are now infected that HPV-associated cervical carcinoma has become the second most common cancer in women worldwide. I'm quiet, but I'm also very popular!

THE INTERVIEWER'S SUMMARY

Human papillomaviruses enter the body through tears or cracks in the skin or the mucosal barrier that separates the body from the outside world. Their targets are the stem cells of the basal epithelium—cells which proliferate slowly to replace the layers of skin or the mucosal epithelium that are continuously being eroded. This small, double-stranded DNA virus enters these cells and uses the cell's machinery to make a small number of copies of its genome—but no complete virions. When these stem cells divide, one of the daughter cells goes back to being a basal stem cell, providing a safe haven for those viral genomes which ended up in that cell. The other daughter cell, carrying the viral genomes it inherited, then begins to differentiate into a keratinocyte or a squamous epithelial cell as it is pushed toward the surface by proliferating basal cells below.

The "hitchhiking" viral genomes are not passive during this journey. In fact, as they ride upward, transcription of viral genes increases and viral proteins are produced. Some of these proteins act to keep the epithelial cell in a proliferative mode, so that a large number of completed virions—viral genomes enclosed in single protein capsid—can be produced. And when the traveling cell reaches the surface, these virions are released. It is in this way that the human papillomavirus uses epithelial cells at different stages of "maturity" as reservoirs for future viral reproduction (basal epithelial stem cells) and as factories for production of new viruses (differentiating epithelial cells). This ingenious scheme even delivers the newly minted viruses to the surface of the body so that they can be spread to new hosts.

The human papillomavirus operates very quietly. After a small burst of viral DNA replication in basal epithelial cells, the virus settles down to replicate its episomal genome at the rate of about one replication cycle per basal cell division. Cleverly, viral RNA is transcribed from only one of the virus' two DNA strands, so the virus usually evades detection by the host's interferon defense system. Moreover, the virus does not kill the cells it infects. In fact, although viral proteins trigger "unscheduled" proliferation of epithelial cells as they travel to the surface of the body, viral proteins protect against the cellular suicide programs that otherwise would destroy these cells. Very few viral proteins are synthesized in basal epithelial cells, and it is difficult for the adaptive immune system to sense that there has been an infection. Moreover, as infected cells journey toward the surface, they leave behind the area where immune system cells are on watch. Consequently, the migrating epithelial cells can devote themselves to producing large amounts of viral proteins without fear of alerting the adaptive immune system. Finally, there are more than a dozen different HPV genotypes that can infect human genital areas, so even when the adaptive immune system gears up to destroy the cells producing one genotype, there are frequently cells infected with other genotypes present which the immune defenses do not recognize.

Most human papillomavirus infections are asymptomatic. Genital warts are the most visible manifestation of an HPV infection, but only a small percentage of persons infected with human papillomavirus have genital warts. Genital warts can be spread to the respiratory tract during oral sex or during the birth of a child to an infected mother—and these warts can be dangerous if untreated. Infection with certain "oncogenic" genotypes of HPV is strongly associated with cervical carcinoma. Although less than 1% of women infected with these genotypes will get cervical cancer, so many people are infected with human papillomavirus that HPV-associated cervical cancer is now the second most common form of cancer in women worldwide.

GENERAL PRINCIPLES

1. Viruses whose DNA is integrated into their target cell's chromosome can be faithfully passed down to daughter cells as the target cell proliferates. In contrast, viruses whose DNA exists in the nucleus of their target cell as an episome run the risk of having their genomes lost as these cells proliferate.

THOUGHT QUESTIONS

1. Discuss what happens between the time the human papillomavirus infects basal epithelial stem cells and the time virus particles are released from the surface of the skin.
2. How does the human papillomavirus avoid being "trapped" in basal epithelial cells as these cells leave the basement membrane, stop proliferating, and dedicate themselves to producing keratin protein?
3. What feature of the virus' "travels" during an infection help it avoid detection by the immune system?

Viruses Which Establish Long-Term Infections

We have "interviewed" six viruses which are able to establish long-term (sometimes lifelong) infections: Hepatitis B, hepatitis C, HTLV-I, HIV-1, herpes simplex, and the human papillomavirus. Because of the strategies these viruses have evolved for their spread—exchange of contaminated blood, intimate physical contact, or vertical transmission from mother to child—long-term infections are essential to provide an extended period during which infection of new hosts can take place. Indeed, for these viruses, an acute infection (with an infectious period of only days or a few weeks) would make persistence in the human population impossible. Although the result is the same—an expanded window of time in which to spread—the strategies used by each of these viruses to establish a long-term infection are quite different.

Hepatitis B and hepatitis C viruses both can establish lifelong, smoldering infections. One of the intriguing features of a chronic hepatitis B infection is that although most liver cells are infected, there is not mass destruction of the liver by killer T cells. Experiments suggest that cytokines produced by immune system cells can interfere with the production of hepatitis B virus—and in some cases can eliminate the virus from infected cells—without killing these cells. Exactly how this happens is not clear, but this might also explain why a chronic hepatitis B infection generally is characterized by periods during which the virus seems to be "hidden" from the immune system (perhaps because viral reproduction is suppressed), and periods during which infectious viruses are actively produced. Indeed, the impression one gets is that the immune system warriors march into the liver and try to deal with the virus-infected cells—but can't quite get the job done. Then, later, when the immune response has "relaxed," more virus is produced, and the cycle repeats. Importantly for the spread of hepatitis B virus, even when the immune system is working its hardest, high levels of infectious virus are present in the blood, being sheltered from immune attack by abundant, empty "decoy" viruses which soak up antiviral antibodies.

An infection with hepatitis C virus also is characterized by "waves" of virus production and relative silence, albeit for a different reason. Hepatitis C virus uses its error-prone polymerase to produce mutant viruses that can stay one step ahead of the adaptive immune system. Consequently, in a hepatitis C infection, there are alternating periods of immune system killing of infected liver cells, and periods of virus production when the immune system must "reboot" to produce weapons appropriate for the mutated hepatitis C virus *du jour*.

When HTLV-I or HIV-1 infects immune system cells, the genetic information of the virus can be integrated into the chromosomes of the infected cells, enabling these viruses to establish a latent, essentially undetectable infection. Once safely tucked away in a cell's chromosome, integrated HTLV-I or HIV-1 proviral DNA can be passed down to daughter cells when the infected cells proliferate. HTLV-I spends most of its time "hiding" in this latent state, and only rarely reactivates to produce more infectious virus. In contrast, although the latent state provides a safe haven for HIV-1, reactivation from these reservoirs to produce more infectious virus is the rule for HIV-1 rather than the exception. As a result, HIV-1 maintains a smoldering, chronic infection in which the virus is engaged in a continuing battle with host defenses.

Whereas HTLV-I and HIV-1 can either productively or latently infect a given immune system cell, herpes simplex virus has devised an elegant solution to the problem of continuous transmission by using two different cell types: epithelial cells in which the virus reproduces efficiently, and nerve cells in which the virus establishes a stealth, latent infection. Also in contrast to HTLV-I and HIV-1, which are able to integrate their genetic information into host chromosomes, the herpes simplex virus genome exists as a "free floating" piece of DNA in the nucleus of latently infected neurons. Some form of trauma, either mental or physical, usually triggers reactivation of herpes simplex virus from its latent state. When the virus reactivates, it reproduces to a limited extent, exits the neuron, and infects nearby epithelial cells to produce new viruses—which are spread by physical contact to new hosts. Alerted by the death of infected epithelial cells, the immune system then wipes out any virus which is not hiding in nerve cells, and the cycle continues.

Basal epithelial stem cells are self-renewing, so infecting them allows the human papillomavirus to establish a long-term reservoir of infection. After an initial burst of replication, the virus settles down to replicate its genome more or less in synchrony with the proliferation of the basal stem cells. Because no viral structural proteins are synthesized in these cells, this is essentially a latent infection—one that is very difficult for the immune system to detect. When basal stem cells divide, the HPV genomes (which are episomal) are randomly allocated to the resulting daughter cells. One of these daughters goes back to being a basal stem cell, providing a long-term reservoir of viral genomes. As the other daughter cell begins to mature to become a keratinized skin cell or a mucosal epithelial cell, viral replication commences, and many new viruses are produced. Then, when these infected cells reach the end of their upward journey to the surface of the skin or vagina, the newly minted viruses are released to infect new human hosts. So in contrast to herpes simplex virus, which uses one cell type (a sensory neuron) for a latent infection and another (an epithelial cell) for its productive infection, the human papillomavirus uses one cell—but at two different maturation stages: a basal epithelial stem cell as a host for latent infection, and a maturing epithelial cell for productive infection.

It is important to realize that many unanswered questions remain about viral latency. For example, virologist don't fully understand how HTLV-I or HIV-1 "decide" to abandon the latent state and produce virions. Likewise, although anyone who suffers from cold sores knows that the reactivation of herpes simplex virus can be triggered by stress or sunburn, exactly how these triggers act to "wake up" the virus is a mystery. Moreover, although it is clear that epithelial cells become permissive for the reproduction of human papillomavirus as these cells mature, it remains to be determined exactly what features of the maturation process allow viral replication. Clearly, there is still a lot to be discovered about how these clever viruses manage to maintain long-term infections in their human hosts.

Part III

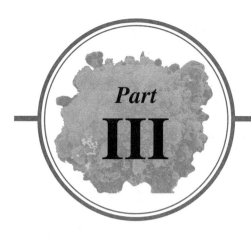

Beyond the Bug Parade

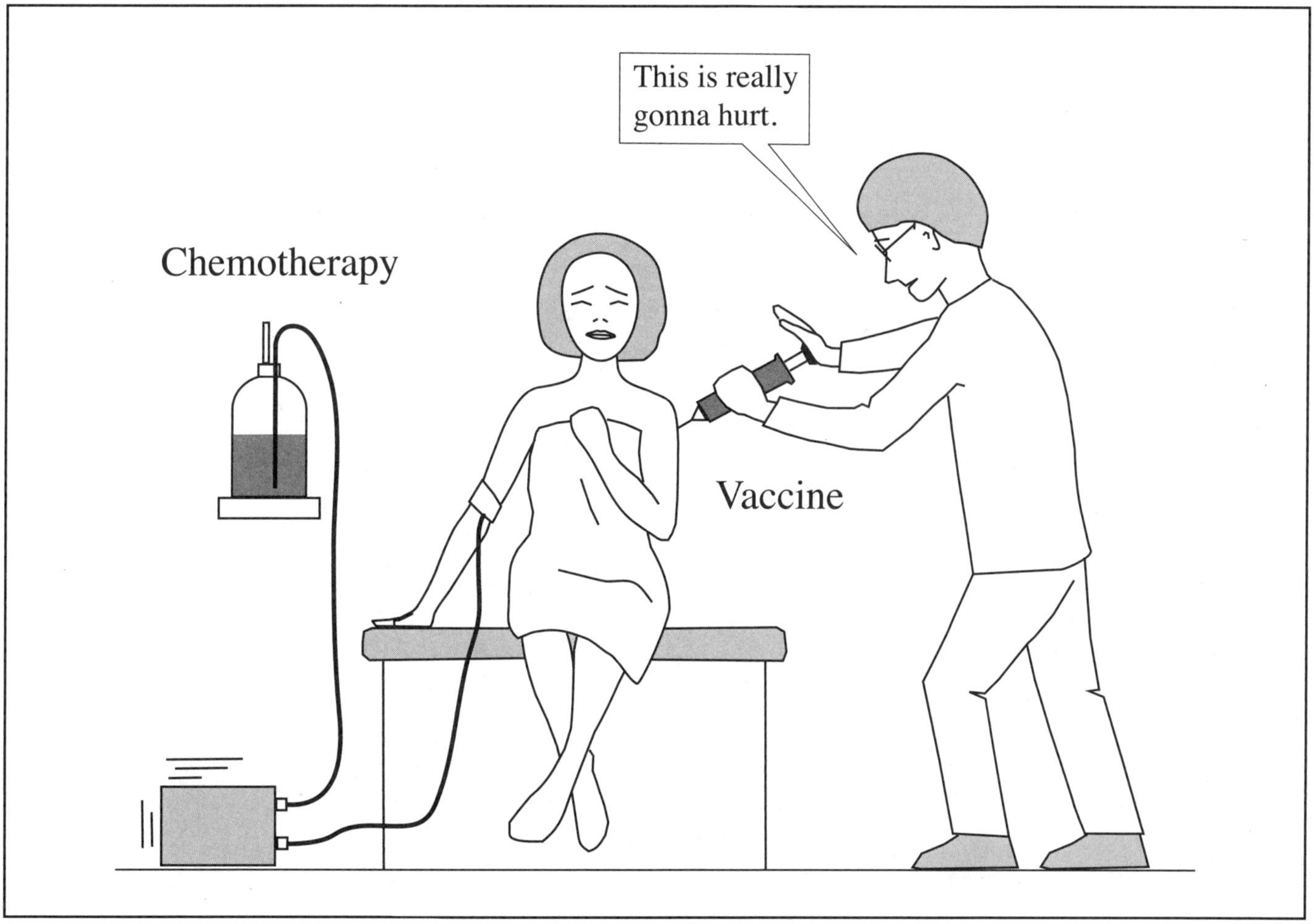

In the next chapter, we will discuss "emerging" viruses — viruses that cause serious disease, and which may become more widespread (and more dangerous!) in the future. Then, in Chapter 17, we will analyze the lifestyles of the four viruses in our Bug Parade which are associated with human cancer, and ask how their ways of "doing business" may contribute to causing this terrible disease.

Although, in many cases, human antiviral defenses are adequate to deal with infecting viruses and prevent serious disease, there are cases when we clearly could use some "outside" help. In the last two chapters, we will discuss the two approaches which have been most successful in augmenting host defenses: vaccinations and the use of antiviral drugs.

Chapter 16

Emerging Viruses

BACKGROUND

There are several definitions of what constitutes an emerging virus. My favorite is a simple one: An emerging virus is one which has only recently become apparent. Of course, viruses have always been emerging, but the ones we really are concerned about are the ones which are emerging now.

WHERE DO EMERGING VIRUSES COME FROM?

Although it is possible that a brand new virus may evolve "from scratch" during our lifetime, it is very unlikely. After all, every virus must learn to solve the problems of how to access and enter its target cells, how to reproduce in these cells, how to evade host defenses, and how to spread to a new host—and all that usually takes a very long time. No, it is much more probable that viruses will "emerge" from existing viruses. This can happen when technological advances make it possible to detect existing human viruses; when changes in human lifestyles allow existing animal viruses to take hold in the human population; when humans "turn over rocks," exposing themselves to viruses which were formerly living in harmony with their animal hosts; and when existing viruses mutate to produce closely related viruses which can infect humans. Indeed, as long as existing viruses are free to mutate, and humans continue to rub shoulders with birds and other animals, new viruses will emerge.

Detection of Previously Cryptic Viruses

Some viruses "emerge" when technology becomes sufficiently advanced to allow their presence to be detected. An excellent example is the discovery of hepatitis C virus. In the late 1960s and early 1970s, hepatitis B virus had been identified as an agent which could cause hepatitis in recipients of blood transfusions. This identification was facilitated by the fact that the blood of infected individuals usually is chock full of decoy as well as infectious hepatitis B virus particles. However when it became clear that all cases of transfusion-related hepatitis could not be attributed to hepatitis B, the search began for other viruses which also might be contaminating the blood supply. Nevertheless, it was not until the late 1980s that advances in biotechnology made it possible to identify hepatitis C virus, and to produce diagnostic reagents which could be used to screen for hepatitis C–contaminated blood. By that time, hepatitis C virus had already infected millions worldwide.

Social Changes Which Lead to Viral Emergence

Viruses can emerge when humans change their lifestyles. Until about 6,000 years ago, humans lived in relatively isolated "tribes." At about that time, cities began to grow up, bringing larger numbers of people into close proximity. This change in lifestyle made it possible for viruses like measles—which have short infectious periods, whose infections confer lifelong immunity, and which only infect humans—to be passed in an unbroken chain from one person to the next. Where measles "emerged from" remains a mystery.

HIV-1 is another excellent example of a virus which emerged due to changes in human lifestyles. Recent analysis suggests that the ancestor of HIV-1 was acquired by humans from chimpanzees around the middle of the twentieth century. HIV-1 does not cause AIDS-like symptoms in chimps, illustrating the important concept that the same virus can cause very different diseases even in species as closely related as chimpanzees and humans. Because HIV-1 has evolved a stable virus-host relationship with chimps, it is likely that HIV-1, or a virus very much like HIV-1, has existed in the chimpanzee population for a relatively long time. If this is true, HIV-1 probably "jumped" from chimps to man much earlier than the twentieth century, but just never caught on in the human population until recently. This view is supported by the fact that HIV-1 is a virus with an "urban lifestyle," and most of Africa only became urbanized in the last 60 years. As more and more Africans moved from the country into larger cities, men left their families and traveled to the cities for employment, increasing the "market" for prostitutes in these urban centers. Indeed, many African men have "second wives" in the city—wives whom they share with other workers. These second wives can act as foci from which HIV-1 infections can spread to workers and then to their families. In addition, urban growth increased the need for long-haul truck transport of various commodities across Africa—and many of the men who drive these trucks visit prostitutes along their routes, facilitating the spread of HIV-1 over great distances. So the urbanization of Africa made it possible for a chimpanzee virus to emerge as a successful human pathogen.

Emergence Due to New Contacts With Animal Viruses

Viruses can emerge when humans "turn over rocks." There are about two dozen non-human primate species which currently are infected with retroviruses related to HIV-1, and virologists must be on guard—because one or more of these viruses might emerge as a human pathogen. If, like HIV-1 and hepatitis C, these emerging viruses were to go undetected for decades, they could contaminate the blood supply, and stealthily infect a significant fraction of the human population.

The clearing of rainforests has brought humans in contact with animal species which rarely were encountered before. The result is exposure to viruses that can infect humans, but for which a human is not the normal host. The most notorious of these are the "hypervirulent" viruses for which humans represent a dead-end infection. These viruses are so lethal that infected humans frequently die before they can pass on the virus. Ebola virus first emerged from an unknown, probably animal host in the African rainforests in 1976. Marburg virus, a close relative of Ebola, was first reported in 1967 when lab workers in Marburg and Frankfurt, Germany, contracted the virus while processing kidneys from infected African green monkeys obtained from Uganda.

Ebola and Marburg are both negative-strand RNA viruses belonging to the filovirus family. They can be passed from human to human by physical contact and perhaps by inhalation, and the majority of infected individuals hemorrhage and die with blood oozing from their gums and nose. Fortunately, quarantine is usually effective in limiting the transmission of these viruses, and outbreaks are sporadic, occurring only every few years. Because these viruses only infect humans sporadically, other natural hosts for Ebola and Marburg must exist—hosts in which infections with these viruses probably cause few or no pathological consequences. To date, however, the natural hosts for Ebola and Marburg have not been identified.

In 1993, in the Four Corners area of northwestern New Mexico, there were a small number of unexplained deaths due to suffocation when fluid accumulated in the lungs of afflicted individuals. The cause of these deaths was quickly determined to be infection with the Sin Nombre (no name) virus. Since that time, several hundred cases of infection with this hantavirus have been documented in many different parts of the United States. Most of these cases occurred in rural areas, and a search revealed that the natural host for this virus is the deer mouse which is widespread in the United States and Canada. These mice can be infected with Sin Nombre virus without obvious pathological consequences, and the virus is most likely spread when mice (or humans) come in contact with virus that has been shed in the urine of infected mice. So the Sin Nombre hantavirus is a good example of a virus which emerged because humans intruded on an established virus-host system. Because the hantavirus quickly kills about 40% of the humans it infects, humans represent a dead-end host.

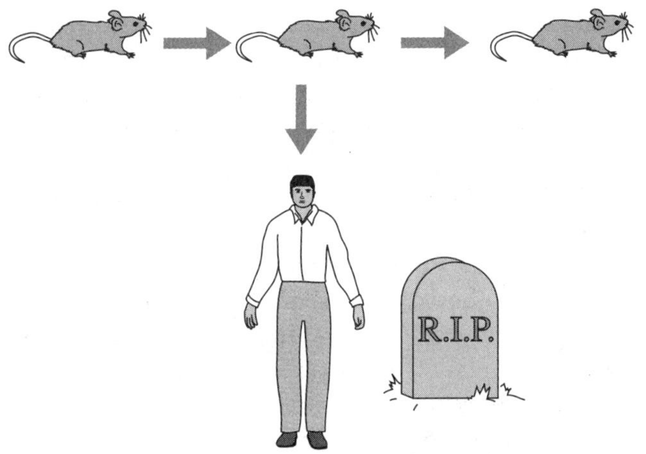

Infections of humans with viruses like the hantavirus, which have birds or animals as their natural hosts, are called **zoonoses**—and most emerging viral infections are zoonoses. Whereas the 12 viruses in our Bug Parade have evolved to live in relative harmony with humans, zoonotic viruses have learned to co-exist with their bird or animal hosts. So for zoonotic viruses, humans are irrelevant for their survival. This means, in principle, that zoonotic viruses can kill as many humans as they wish without endangering their ability to be maintained in the bird or animal population.

In the summer of 1999, the West Nile virus killed seven people in the New York City area. This virus is fairly common in parts of Africa, Europe, and Asia, but only recently "emerged" in the United States. Birds are the natural hosts for the West Nile virus, and the "vector" which transmits the virus from bird to bird is the mosquito. Humans can also be infected with the West Nile virus when they are bitten by infected mosquitos, but human infection is probably a dead-end for this virus: The amount of West Nile virus in the blood of an infected human usually is too small to be efficiently transmitted by a mosquito to another host.

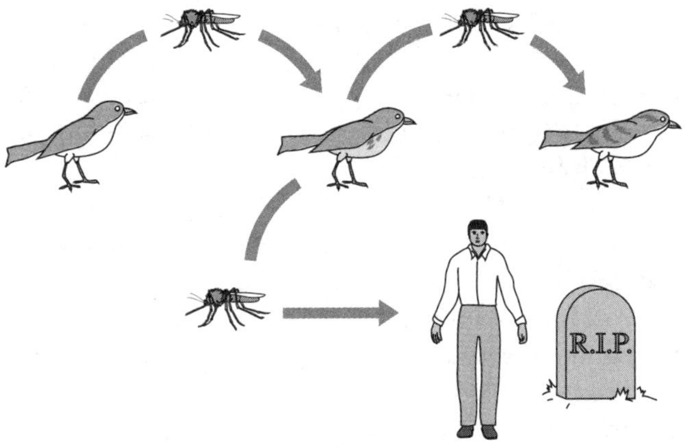

The "rock" that was turned over to allow the emergence of the West Nile virus in the United States was likely the transport of infected mosquitos or birds to this country from another part of the world, probably the Middle East. Indeed, world travel is responsible for the emergence of many viruses in new locations. More than a million passengers travel internationally each year, making spread by human carriers commonplace. And because air travel is so rapid, infected individuals can reach their destinations before any symptoms appear.

West Nile is a positive-strand RNA virus, and is a member of the same family as hepatitis C virus—the flavivirus family. It can infect more than 60 bird species, including American robins, crows, blue jays, and sparrows. Having birds as hosts is a great idea for a virus, because birds, like humans, are big travelers, and can spread the virus over large geographic areas. Moreover, pools of water in discarded tires make ideal breeding grounds for virus-infected mosquitos, and these tires are routinely shipped across the country for recycling. Indeed, it only took about four years for the West Nile virus to spread all the way from New York City to the West Coast.

Between 1999 and 2010, almost two million Americans were infected with West Nile virus, and about 1,300 of these people died as a result of the infection. Fortunately, by keeping the mosquito population under control, the number of West Nile infections can be kept to a minimum. For example, once the vector was identified, increased spraying for mosquitos in the New York area helped trim the number of deaths attributed to the West Nile virus from seven, in the summer of 1999, to only one in the summer of 2000.

Variants of Existing Viruses

Viruses can emerge when existing viruses mutate. Successful viruses co-evolve with their hosts and settle down into a lifestyle that allows them to spread efficiently within the host population. However, the evolution of viruses is driven by mutation, and viral mutation does not cease even when viruses become successful pathogens. Many viruses use error-prone polymerases to replicate their genetic information. In extreme cases (e.g., hepatitis C virus and HIV-1) this results in a mutation rate so high that essentially every new virus produced will be a mutant. Viruses such as influenza, which have segmented genomes, can also mutate by exchanging gene segments with closely related viruses that infect other species. In addition, when retroviruses like HIV-1 and HTLV-I integrate their genetic information into the genome of a host cell, it is relatively easy for these viruses to "steal" cellular genes or to recombine with other retroviruses which may have infected the same cell. The result of such events can be mutated retroviruses with altered properties.

Mutations in viral genomes also can change the host range of a virus, allowing the virus to "jump" to a new species. We have already discussed how influenza virus can jump from birds to humans when the viral genome mutates by the reassortment of gene segments during the infection of pigs—an animal which hosts both bird and human flu viruses. Another example of this type of "adaptation" is the SARS virus (SARS-CoV).

The virus that causes severe acute respiratory syndrome (SARS) is a large, positive strand RNA virus which is a member of the coronavirus family. This family has more than 20 variants which can infect at least 30 different animal species, including humans. Indeed, about 15% of all cases of the common cold are caused by two of these variants. In contrast to these two human coronaviruses, the precursor of the SARS virus normally infects bats, where it causes no apparent disease. Further, there is no indication that the bat coronavirus can spread directly from bats to humans. So how did the bat coronavirus manage to infect humans? The story is quite interesting.

In China, there are "wet" markets where live animals are sold for food, and one animal sold in these markets is the horseshoe bat. Also sold in these wet markets are palm civets—small, cat-like animals. Genetic analysis indicates that, in the crowded environment of the market, bats infected with coronavirus infected palm civets. The coronavirus RNA polymerase is quite error-prone, and as the virus reproduced in the civet, its S protein, which is involved in binding to its target cell, mutated so that it could bind more tightly to receptors on human cells—making human infection by this mutated version of the original bat virus more efficient. Food handlers in the market then became infected with this new virus, and the SARS epidemic began. During the period from 2002–2003, this mutated virus spread to 33 countries, infected more than 8,000 people, and killed 813 of them. Since 2004, the SARS coronavirus has essentially disappeared from the human population—and virologists don't know why.

The ability of viruses to rapidly change their genomes also makes it possible for them to become more virulent. In 1918, mutational events occurred which changed influenza from a cold virus into a virus that killed about 20 million people. This virus was so virulent that there are reports of previously healthy soldiers who collapsed, and then died only one day later.

The 1918 viral genome has now been recovered from lung tissue preserved from individuals who died from the infection. Sequence analysis suggests that the virulent 1918 influenza H1N1 strain probably was an avian virus that adapted (mutated) to infect humans. A current hypothesis is that the original jump from bird to human occurred early in 1918, and was a strain with low pathogenicity. Then, as this strain circulated in the human population, antigenic drift produced a killer virus which, instead of being confined to the airway, was capable of establishing a deadly, systemic infection. Experiments with mice indicate that the concerted action of several of the proteins encoded by the 1918 virus is responsible for its **virulence**. One of these is the viral hemagglutinin. Whereas the hemagglutinin protein of seasonal influenza must be cleaved by an enzyme present in the respiratory tract, analysis of the hemagglutinin protein of the 1918 virus suggests that proteases which are more widely distributed in the body may be able to perform this activating cleavage—although the postulated "promiscuous" proteases have not been identified.

THE DANGERS OF EMERGING VIRUSES

So far, public attention has focused mainly on emerging viruses like Ebola or hantavirus, which seem to strike at random and which kill a large fraction of the humans they infect. They make good headlines. However, to my mind, these hypervirulent viruses are not the most dangerous ones. First, these viruses are not easily passed from human to human, and certainly not by casual contact. Ebola virus, for example, can be spread by infected humans, but when this happens, the infected individual is usually bleeding from the nose and gums—hardly someone who would be contacted "casually." As a result, quarantine measures are quite effective in containing outbreaks of viruses like Ebola and Marburg. In addition, because hypervirulent viruses kill humans so efficiently, their natural hosts must be non-human—hosts in which they have evolved to do little damage and to spread easily. This means that so long as humans keep their distance from these diseased, non-human hosts, it is unlikely that we will be infected by these zoonotic viruses. For instance, the hantavirus' natural host is the mouse, and the virus spreads when mice snuffle surfaces on which infected mice have urinated. Since humans are unlikely to be snuffling the urine of other humans infected with the hantavirus, keeping the mouse population under control should prevent most hantavirus infections.

No, my idea of a really dangerous virus is not one of these hypervirulent viruses we read so much about. Indeed, the killer virus we should fear (let's call it the "Andromeda Strain") will likely have the following characteristics. First, the Andromeda Strain will have humans as its natural host (or at least one of its natural hosts). A virus which relies on an animal or bird reservoir, and for which humans represent only a dead-end infection, would be too easy to deal with—as we have already discussed.

Second, to allow for the infection of a large fraction of the human population, the Andromeda Strain will be easily spread by casual contact. Such a virus, for example, might be spread by the respiratory route. That way, you wouldn't even have to touch the infected person to become infected—you'd just have to be nearby when he coughed or sneezed.

Third, although the virus will kill a high proportion of the people it infects, the Andromeda Strain will kill its hosts only after a long, relatively asymptomatic infection. This feature will give the virus a long window of infection, during which the virus can be transmitted to other hosts.

I think it's unlikely that the Andromeda Strain already exists somewhere on earth. With people now living all over the globe, and with rapid, world-wide transportation to disseminate the virus, the chance is great that at least some humans would have turned over the "rock" under which such a virus was hiding. This means that the Andromeda Strain probably will evolve from a present-day virus which mutates rapidly.

Could It Be HIV-1?

The urbanization of the twentieth century has already produced a near miss for the Andromeda Strain: HIV-1. This virus has most of the qualities we fear in a virus. HIV-1 eventually kills almost all of the humans it infects. It has a long, mostly asymptomatic infectious period made possible in part by its high mutation rate, and partly because it uses the host's immune system to its own advantage and then destroys it. Currently, about 0.5% of the world population is infected with HIV-1, and the virus continues to spread. However, HIV-1 is not the Andromeda Strain. The reason, of course, is that it is not spread by casual contact. Usually, you have to "do something" to contract this virus: have intimate physical contact with someone who is carrying the virus, share needles with someone who is infected, etc. Indeed, in the United States, where the blood supply is screened, with the exception of being the child of an infected mother, you should be able to avoid infection with HIV-1 completely.

Imagine, however, what could happen if HIV-1 somehow mutated to be spread efficiently by the respiratory or fecal-oral route. Then HIV-1 would be the Andromeda Strain! In fact, this may be the greatest danger that HIV-1 poses: the possibility that this virus may mutate to be spread by casual contact. After all, we already have discussed several examples of viruses which can mutate to change their mode of transmission. Human strains of influenza virus are spread by the respiratory route, but antigenic drift can produce a strain which, at least in ducks, is spread by the fecal-oral route. Likewise, some strains of human adenovirus are spread by the respiratory route, whereas other, closely related strains are spread by the fecal-oral route. Thankfully, changing the mode of transmission from intimate physical contact to the respiratory route is probably much more difficult than switching from respiratory to fecal-oral. However, HIV-1 mutates very rapidly, huge numbers of new (mutated) viruses are produced each day in every infected individual, and millions of humans are infected. So the emergence of strains of HIV-1 which can be transmitted by casual contact is not something that any virologist would rule out. And if this were to happen, humans would be in deep, deep trouble.

Could It Be Influenza?

Although the 1918 flu virus killed millions, it was not the Andromeda Strain. The world's population in those days was about two billion, so only about 1% of the population died during the influenza pandemic. This was partly because the less virulent strain which circulated earlier in the year probably immunized many against the killer strain that came later. But more importantly, the 1918 strain of influenza A was so virulent that most who were infected died before they could distribute the virus widely to others. Consequently, although the 1918 flu had some of the properties we would predict for the Andromeda Strain—evolution from an existing human virus which has a high mutation rate and which is spread by casual contact—still, it lacked one very important attribute: a long, relatively asymptomatic contagious period. It was probably for this reason that the 1918 influenza strain "burned itself out" after only two years.

So the 1918 flu was not the Andromeda Strain—but it came very close. A vast pool of influenza A virus genes exists in birds and other animals (e.g., pigs), and influenza can easily tap into this gene pool to create new viral strains. Imagine what might happen, for example, if influenza virus mutated to be as deadly as the 1918 strain, but killed its victims more slowly, say over a relatively asymptomatic period of a few months. This would give infected individuals an extended opportunity to spread the virus by coughing and sneezing. Of course, to pull this off, the virus would have to learn to evade the immune system during this long infectious period—and that's not a trivial problem for a virus to solve. However, given the large reservoir of influenza virus sequences, and influenza's propensity for antigenic shift and drift, we must be vigilant lest a new variant of influenza evolves unnoticed—a virus which might indeed have the properties of the Andromeda Strain.

Could It Be the SARS Coronavirus?

The SARS coronavirus certainly has some of the attributes which we might expect in the Andromeda Strain. It is spread primarily by the respiratory route, and it is extremely contagious. During the 2002–2003 epidemic, a single patient in one hospital infected 138 other patients and healthcare workers. The SARS coronavirus can establish a systemic infection, and is quite lethal, with an overall mortality rate of about 10%. Sadly, many of the casualties of this epidemic were hospital workers. In fact, the physician who first identified the disease and sounded the alarm, Dr. Carlo Urbani, was one of its early victims.

At the beginning of an infection, SARS can easily be confused with other diseases because most patients present with common, flu-like symptoms (fever, chills, cough, diarrhea, etc.). This makes it difficult to make an early diagnosis. Moreover, the contagious period generally lasts for three to four weeks, giving the virus plenty of time to spread to new hosts.

Although SARS has some of the features of the Andromeda Strain, the path from the animal reservoir in bats to humans is rather easily interrupted. In addition, the contagious period does not begin until a patient begins experiencing the serious symptoms of a lower respiratory tract infection (e.g., pneumonia). Indeed, it was this feature which made it possible to control the spread of the SARS coronavirus by identifying potentially infected patients and quarantining them.

So far, there have been two major SARS outbreaks: the first in 2002 to early 2003, and a second from late 2003 until early 2004. Genetic analysis of viruses from these outbreaks showed that they have major differences. Consequently, they most likely arose from two separate events in which an animal coronavirus "jumped" into the human population. This means that although such species-crossings appear to be rare, they are not unique, one-time events—and this, of course, raises the possibility of additional SARS epidemics. Moreover, it is not known why the two earlier epidemics died out, so there is always the possibility that a new strain will emerge which includes mutations that will allow it to persist in humans and be less easy to quarantine. That is a very scary possibility.

Although HIV-1, influenza A, or the SARS coronavirus do not qualify as the Andromeda Strain, they come frighteningly close. From our experiences with these three near misses, it is clear that we must take emerging viruses very seriously, and that virologists must try to evaluate the pathogenic potential of such viruses early after their emergence.

GENERAL PRINCIPLES

1. Viruses can emerge when technological advances make it possible to detect existing human viruses; when changes in human lifestyles allow existing animal viruses to take hold in the human population; when humans "turn over rocks," exposing themselves to viruses which were formerly living in harmony with their animal hosts; and when existing viruses mutate to produce closely related viruses which can infect humans.
2. The same virus can cause very different diseases even in species as closely related as chimpanzees and humans.
3. Infections of humans with viruses which have birds or animals as their natural hosts are called zoonoses — and most emerging viral infections are zoonoses.
4. To be truly dangerous, a virus would likely have the following properties: It would have humans as a natural host; it would be easily spread by casual contact; and it would kill humans only after a long, relatively asymptomatic period during which it could spread to other humans. Such a virus would probably arise from a present-day virus which has a high mutation rate.

Chapter 17

Virus-Associated Cancer

BACKGROUND

Cancer results when two types of cellular control systems are corrupted: systems that are designed to promote cell proliferation, and safeguard systems which are supposed to protect against "irresponsible" cell growth. These control systems are composed of genes and the proteins they encode, and these systems can malfunction when the structure of the genes which comprise them is altered—usually by mutations in cellular DNA. These changes can lead to reduced or increased expression of genes or to the production of altered proteins. It is estimated that about five control-system mutations are required to produce most common cancers. Because it usually takes a long time to accumulate these mutations, cancer is a disease that normally appears late in life.

Although these cancer-causing mutations occur "spontaneously" throughout a person's life (e.g., because of mistakes made in copying cellular DNA), there are factors which can accelerate the rate of mutation. For example, if a person smokes cigarettes, eats a fatty diet, or is exposed to high levels of radiation, the rate at which these mutations accrue accelerates. In addition to these "environmental" factors, some viruses encode proteins that can interfere with the proper functioning of these same cellular control systems. Consequently, infection with one of these viruses can contribute to an infected cell becoming cancerous. I say "contribute" because it is important to note that no virus "causes" cancer. A viral infection only acts as a risk factor—just like cigarette smoking. Indeed, a hallmark of a cancer-associated virus is that relatively few people who are infected with the virus will get cancer, yet most of those who do will have evidence of a virus infection. It also is essential to realize that most viral infections have nothing to do with cancer. Only a select few viruses—the "oncogenic viruses"—play any role in cancer.

In this chapter, we will discuss the four oncogenic viruses in our Bug Parade, all of which are associated with human cancer: hepatitis B virus, hepatitis C virus, the human T cell leukemia virus (HTLV-I), and the human papillomavirus (HPV). You will remember that each of these viruses causes a long-term, often lifelong, infection. This is a "minimum requirement" for a virus to be a tumor virus—because it usually takes years to accumulate the genetic alterations which cause a cell to become cancerous. Although virologists have been working for a long time to discover the mechanisms by which viruses help cause cancer, much about virus-associated cancer remains mysterious.

HEPATITIS B VIRUS

Approximately 20% of long-term, hepatitis B carriers can be expected to contract liver cancer (hepatocellular carcinoma), and about one million people die of hepatitis B–associated liver cancer each year. Exactly how this virus acts as a risk factor for cancer is not well understood, although it is likely that several facets of the virus' lifestyle play a role. One of the primary duties of liver cells is to detoxify potentially damaging chemicals which either enter the blood from outside the body, or which originate within the body as toxic byproducts of normal cellular metabolism. Many of these toxins (the "genotoxins") can directly or indirectly damage cellular DNA, and although liver cells are stocked with enzymes which can detoxify these chemicals, sometimes these detoxification systems can be overloaded. When this happens, liver cells become targets of the very genotoxins they normally are able to protect against.

Usually, liver cells are not proliferating, and in a resting state they generally have time to repair damage inflicted by genotoxins. However, in a hepatitis B–infected liver, cells must proliferate to replace those which have been killed by the immune response to the infection. And this "extra proliferation" increases the risk that these liver cells will divide before DNA damage can be repaired. So the combination of being constantly exposed to genotoxins while being forced to proliferate may predispose cells in hepatitis B–infected livers to cancer-causing mutations.

More than 90% of hepatitis B–associated liver cancers contain at least partial hepatitis B genomes that have somehow been inserted into a cellular chromosome. This suggests that the virus does something to these cells which actively contributes to the cancer-causing process. Most research designed to discover which hepatitis B viral functions might be involved in cancer has centered on the viral X protein. This protein is essential for viral reproduction and spread within infected hosts. Although the exact role of the X protein in hepatitis B infections remains unclear, it has been shown that this protein can inactivate the important tumor suppressor, p53.

When a cell with unrepaired DNA damage begins to proliferate, that cell usually is stopped in its tracks by the action of the p53 protein. If the DNA damage is minor and can be repaired, proliferation is halted until the repair has been completed. On the other hand, if DNA damage is extensive, the p53 protein triggers the cell to commit suicide. It is for this reason that the p53 protein is often referred to as the "guardian of the genome." Because the X protein has been shown to inactivate p53, cells infected with hepatitis B virus are likely to be more susceptible to cancer-causing mutations. Still, it usually takes many years for these mutations to accumulate, and hepatitis B-associated liver tumors generally arise 20 to 50 years post infection.

HUMAN PAPILLOMAVIRUS

One of the viruses whose mechanism of oncogenesis is best understood is the human papillomavirus (HPV). The reason for this is that similar, small DNA viruses (e.g., SV40 and polyoma viruses) are associated with cancer in animals, and these animal models have been especially useful in teasing out oncogenic mechanisms. Although HPV is associated with cervical cancer in humans, of the many human papillomavirus types, only about a dozen are classified as oncogenic. HPV-16 and HPV-18 are most consistently found in cervical cancers, with HPV-31, HPV-33, and HPV-45 being found less frequently. Clearly, human papillomavirus infections alone do not "cause" cervical cancer, because less than 1% of all women infected with genital HPV will ever suffer from this disease. However, in over 95% of the cervical cancers that have been carefully examined, DNA of one or more of the oncogenic HPV types has been detected. So although an HPV infection is not sufficient to cause cervical cancer, in the vast majority of cases, it is a necessary element in the development of this disease.

HPV encodes two proteins, E6 and E7, which are essential for keeping infected epithelial cells in the "proliferative mode" required for viral infection. Although all human papillomaviruses express these proteins, the E6 and E7 proteins encoded by oncogenic types of HPV differ from their non-oncogenic counterparts in at least two respects. First, the non-oncogenic E6 and E7 proteins are relatively "weak," and usually stimulate only a small amount of "extra" cell proliferation—just enough to allow the virus to reproduce. In contrast, the oncogenic E6 and E7 proteins are more disruptive of the cell cycle than are their non-oncogenic counterparts. For example, the oncogenic E7 protein binds more strongly to the cellular pRB protein than does the non-oncogenic version of E7, and as a result, gives cells a bigger "kick" to cause them to proliferate. It's as if the oncogenic human papillomaviruses want to make absolutely sure they don't get trapped within an epithelial cell which is not proliferating. And to avoid the consequences of this bigger kick, the oncogenic E6 protein promotes the degradation of the cellular p53 tumor suppressor—a protein which normally would trigger suicide (apoptosis) in a cell whose DNA replication is really out of sync.

So the oncogenic versions of the HPV E6 and E7 proteins are "stronger" than their non-oncogenic counterparts, but this is only half the story. In HPV-infected cells, the viral genomes usually exist as circular, double-stranded DNA molecules which float free in the nucleus of an infected cell, unassociated with host chromosomes. This is because, unlike HIV-1 or HTLV-I, the human papillomavirus does not encode an integrase protein which can facilitate the precise integration of viral DNA into cellular chromosomes. Nevertheless, on rare occasions, HPV DNA can be integrated into the cellular genome. This "unaided" integration of HPV DNA is a random event, which probably happens during the repair of damaged cellular DNA. Because integration of HPV DNA is imprecise, viral genes usually are disrupted or lost when the viral DNA is pasted into a cellular chromosome. Of course, this loss of viral genetic information effectively inactivates the virus, so integration of its DNA into the genome of a cell is a dead-end event for a human papillomavirus.

Although unaided integration usually results in the loss of some HPV genes, the genes encoding the E6 and E7 proteins of the oncogenic HPV genotypes are almost always found intact in cervical cancer cells. Moreover, integration of these viral genes can result in the increased expression of the E6 and E7 proteins. In some instances, the gene for a viral protein (E2), which normally helps keep expression of E6 and E7 at relatively low levels, is deleted during the integration event. In other cases, signals within the mRNAs that code for E6 and E7, which normally would shorten the lifetimes of these mRNAs, are clipped off during integration, resulting in more-stable E6 and E7 mRNAs and correspondingly higher E6 and E7 protein levels.

As long as the viral genome is not integrated into the chromosome of an infected cell, the added strength of the oncogenic E6 and E7 proteins appears to be of little consequence. However, when the genes for oncogenic E6 and E7 are fortuitously integrated so that these proteins are expressed at unnaturally high levels, things can get ugly. Now the cell is strongly driven to mutate by the overexpressed (and more powerful) E6 and E7 proteins. Moreover, the tumor suppressor protein, p53, which normally would keep mutant cells from becoming cancerous, has been eliminated by the action of E6. So it is believed that infection with an oncogenic human papillomavirus facilitates the evolution of cancer cells when viral genes, which are absolutely required for viral reproduction, are expressed "unnaturally" from an integrated viral genome. The proteins encoded by these viral genes function to increase the rate at which infected cells mutate, and to disable mechanisms which normally protect cells from these mutations.

Fortunately, integration of an oncogenic HPV genome in such a way as to deregulate E6 and E7 expression is a rare event. And that's one reason why so few women infected with HPV ever get cervical cancer. In addition, it takes more than the destruction of one tumor suppressor protein (e.g., p53) to create a human cancer cell: Additional mutations in cellular genes must occur before a cell with integrated HPV genes can become cancerous. That's why cervical carcinomas usually arise decades after the initial HPV infection. Cigarette smoking and immunosuppression (e.g., co-infection with HIV-1) are two risk factors which, together with an oncogenic HPV infection, help create an environment permissive for cervical cancer.

During their "pre-cancerous" period, HPV-infected cells may accumulate mutations and progress through stages in which they become increasingly more cancer-like. In the final stage of this progression, the mutating cells may "learn" the deadliest trick of all: how to break through the basement membrane and metastasize to other parts of the body.

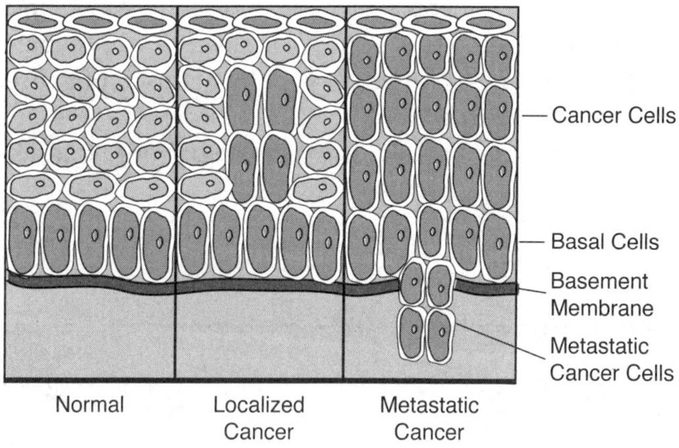

Although oncogenic human papillomaviruses can infect several other areas of the female reproductive tract (e.g., the vulva and the vagina), most HPV-associated cancers occur in the "transformation" zone of the cervix—the area where the epithelium changes from the multilayered, squamous epithelium of the vagina to the single-layered epithelium of the endocervix. Why this region is a hot spot for cervical cancer is a complete mystery. Nevertheless, because cervical cancer is usually limited to this one area, and because cervical carcinoma generally develops in stages over a period of years, Pap smears taken from the transformation zone can be extremely valuable in diagnosing cervical cancer in the early stages.

It is important to note that although oncogenic HPV types can cause warts on the exterior genitals, most HPV-associated genital warts actually are caused by non-oncogenic HPV genotypes (usually HPV-6 or HPV-11). However, because it is common for individuals to be infected with more than one HPV genotype at a time, the presence of genital warts can signal a possible infection with oncogenic genotypes.

Of course, men also can be infected with oncogenic HPV genotypes (that's how women get infected, right?). However, HPV-associated penile cancer is relatively rare—perhaps because the penis has no transformation zone analogous to that of the uterine cervix. In this regard, it is interesting that a similar transformation zone does exist in the respiratory tract where the multilayered, stratified squamous epithelium of the vocal cords meets the pseudo-stratified epithelium that lines the trachea. It is this area which is the primary target for respiratory HPV infections.

HEPATITIS C VIRUS

About 10% of patients with hepatitis C–induced cirrhosis of the liver eventually suffer from liver cancer (hepatocellular carcinoma), which usually arises about three decades post infection. Two viral proteins (core and NS5A) have been shown to inactivate p53 or to interfere with its function, and transgenic mice which have been engineered to express the viral core protein get liver cancer. However, it is not known for certain how a hepatitis C infection predisposes humans to liver cancer.

HTLV-I

Most people who are infected with HTLV-I experience very few adverse consequences, but about 2% of those infected will eventually contract a blood-cell cancer—adult T cell leukemia (ATL). This type of leukemia usually is diagnosed when patients are in their 40s or 50s. Although the details are far from clear, it is likely that ATL results because control systems become dysregulated during a chronic HTLV-I infection when T cells are stimulated over and over by the viral Tax protein.

GENERAL PRINCIPLES

1. Some viruses encode proteins that can interfere with the proper functioning of cellular control systems. Consequently, infection with one of these "oncogenic" viruses can contribute to an infected cell becoming cancerous.
2. No virus "causes" cancer. A viral infection only acts as a risk factor—just like cigarette smoking.
3. Only viruses which cause long-term infections are associated with an increase in cancer susceptibility. However, not all viruses which cause long-term infections are associated with human cancer (e.g., herpes simplex virus).
4. A hallmark of a cancer-associated virus is that relatively few people who are infected with the virus will get cancer, yet most of those who do will have evidence of a virus infection.

Chapter 18

Vaccines

BACKGROUND

The current AIDS epidemic is especially worrisome. Not only does this virus already have many properties of the Andromeda Strain, it mutates rapidly, so it has the potential to acquire additional, deadly characteristics. Of course, this danger could be eliminated if humans were to modify their behavior to prevent its spread. However, modifying human behavior is something which is notoriously difficult to do. Vaccines are incredibly powerful weapons for controlling infectious diseases. Indeed, smallpox virus, which once killed millions, has been eradicated from the earth through the use of a vaccine. One of the great advantages of a vaccination is that, for the vaccine to be effective, the recipient does not have to make any changes in hygiene or lifestyle: He simply has to show up and be vaccinated.

Many virologists believe that the only way to control the spread of HIV-1 is to produce a vaccine which will protect against this virus. However, making an AIDS vaccine has proven to be a daunting task. Indeed, because HIV-1 is a showcase for the problems which can arise in producing an effective vaccine, we will use HIV-1 as our "model" virus in this chapter. As we discuss strategies currently used to make vaccines, we will evaluate whether any of these approaches might be suitable for producing a safe and effective AIDS vaccine. I think you'll come to appreciate that this is really a knotty problem—one which may turn out not to have a solution.

MEMORY CELLS

When a person is first infected by a virus, that person's immune system usually produces memory B cells and T cells. In comparison to the B and T cells of a person who has never been infected by the virus, these memory cells are more numerous and more easily mobilized. Consequently, they often can protect against reinfection by destroying the virus they remember before it can become established. For this reason, the goal of any vaccination is to generate memory B and/or T cells—cells which can provide protection if the vaccinated person is subsequently exposed to the "real thing." To create a vaccine that will produce memory cells, immunologists employ strategies that are similar to those used by our military in its "war games." To prepare our troops for battle, the planners of these games strive to present them with as realistic a view of war as is possible—without putting them in great danger. Likewise, vaccine developers try to design a vaccine that will give the immune system a good look at an invader (e.g., HIV-1) without endangering the recipient of the vaccine.

NONINFECTIOUS VACCINES

One strategy used for memory cell production involves vaccination with an agent that cannot infect the vaccine recipient. An example of such a "noninfectious" vaccine is the killed polio virus vaccine devised by Dr. Jonas Salk. To produce this vaccine, Salk and his co-workers first grew large quantities of polio virus in the laboratory by infecting monkey cells growing in petri dishes. Then they treated this virus with chemicals such as formaldehyde that "glue" viral proteins together, and used this disabled virus as a vaccine. The beauty of this approach is that Salk's killed virus looks just fine to the immune system, but it can't cause an infection because of the formaldehyde treatment. Noninfectious vaccines are quite common. For example, the flu vaccine we get each year is a killed virus vaccine.

One important characteristic of vaccines made from "killed" viruses is that, although the chemical treatment will certainly disable most of the viruses, it is impossible to guarantee that it will kill them all. This is not a problem if the disease in question is spread by casual contact, and if a relatively large fraction of the population is at risk of infection—as was the case with polio in the 1950s. However, for a virus like HIV-1, which in most cases can be avoided, a killed virus vaccine that has even a remote possibility of causing disease would be unacceptable.

Another strategy for making noninfectious vaccines is to use genetic engineering to produce one or a few viral proteins in the laboratory, and to use these proteins as a vaccine. Such a vaccine is called a "subunit" vaccine, and this technique is currently used to make very effective vaccines against hepatitis B virus and the human papillomavirus. Subunit vaccines have the advantage that, because only one or a few viral proteins are present, there is no possibility that the vaccination will result in an infection. On the other hand, the limited number of viral proteins included in a subunit vaccine gives the immune system fewer targets to focus on, and this paucity of targets can be problematic when vaccinating against viruses that have high mutation rates.

Noninfectious vaccines are very effective in fooling the immune system into producing memory B cells—which produce antibodies that can tag the invader for destruction. However, killer T cells are designed to detect and destroy living, virus-infected cells. Consequently, a noninfectious vaccine will not mobilize memory killer T cells.

Whether or not the lack of killer T cells will be a problem in controlling a viral infection is hard to predict. For example, the Salk polio vaccine works well, so clearly a noninfectious vaccine that causes the production of virus-specific antibodies can defend against a polio virus infection. On the other hand, killed virus vaccines for measles and mumps were real duds. So it just depends on the virus. Unfortunately, there is a strong feeling among immunologists that memory killer T cells will be essential to resist an HIV-1 infection. If true, a noninfectious vaccine would not protect against the AIDS virus.

ATTENUATED VIRUS VACCINES

The famous Sabin polio vaccine was produced using another strategy. To make this "attenuated" vaccine, Dr. Sabin grew the polio virus for many generations in monkey kidney cells rather than in human nerve cells—the cell type this virus usually infects. For reasons that are not well understood, growing a virus in the "wrong" host can produce mutations in the virus that weaken it. In this case, Sabin's experiments resulted in three strains of polio virus that could infect a vaccine recipient, but which were so weak (attenuated) that they did not cause disease in healthy people. These three strains were then combined to make the Sabin vaccine. A similar strategy has been used to produce the attenuated mumps and measles vaccines that are in common use today.

Preparation of an attenuated virus vaccine involves a bit of magic, because attenuated viruses must walk a fine line: They must be strong enough to produce a vigorous immune response, yet weak enough not to cause disease. To determine whether the attenuation strategy has worked, the vaccine can be tested on animals, assuming an appropriate animal can be found. But ultimately, the vaccine must be tested on human "volunteers." Interestingly, by the time Sabin was ready to test his vaccine, most Americans had already been vaccinated with the Salk vaccine—and, of course, it wouldn't make sense to test his vaccine on Americans who had already been vaccinated against polio. So Sabin had to look elsewhere. And where do you think Sabin went to find his volunteers? To Russia! No, I'm not kidding you. During the Cold War, Sabin tested his vaccine in Russia—a country whose people had not yet been vaccinated against polio.

Vaccines made from attenuated (weakened) microbes have an important feature: They usually provide lifelong immunity. This is in contrast to noninfectious vaccines which frequently require periodic booster vaccinations to maintain protective antibody levels. Moreover, because an attenuated virus can infect cells, an attenuated virus vaccine can produce memory killer T cells. Of course, the fact that an attenuated virus actually infects the recipient raises several safety issues. First, although a healthy immune system usually will wipe out an attenuated virus before it can cause serious disease, this may not be the case for an individual whose immune system has been weakened. In fact, if a healthy person who has just been inoculated with an attenuated virus vaccine passes the virus on to someone who is immunosuppressed (e.g., due to cancer chemotherapy), there is a chance that the immunosuppressed person's immune system will be too weak to fight off the attenuated virus. Also, because the attenuated virus will multiply to some extent in a healthy vaccine recipient, there is a slight possibility that the crippled virus will mutate, and again become strong enough to cause disease. The probability that such a mutation will occur is low, but even the Sabin attenuated virus vaccine has caused a few cases of polio in healthy recipients due to mutations that restored the strength of one of the attenuated viruses used in this vaccine. Because the AIDS virus has an extremely high mutation rate, this could be a real concern if HIV-1 were attenuated and used as a vaccine.

CARRIER VACCINES

To efficiently produce memory killer T cells, a vaccine must actually be able to infect cells. However, safety concerns dictate that an infectious, attenuated AIDS virus would be unacceptable because of the risk of mutation. To try to resolve this dilemma, immunologists are trying new approaches to vaccine design.

One strategy involves using a virus or a bacterium that doesn't cause disease (e.g., the vaccinia virus used for decades to vaccinate against smallpox) to "carry" one or more HIV-1 genes into cells. In this way, carrier-infected cells could be tricked into producing these HIV-1 proteins in addition to the carrier's own proteins, and memory killer T cells that recognize the HIV-1 proteins could be produced. Most importantly, although a carrier vaccine infects cells and should produce memory killer T cells, there is no chance that the vaccine would cause AIDS—because only a few HIV-1 genes are "carried" by the vaccine.

It would seem that a carrier virus vaccine would be perfect for protecting the general population against HIV-1, and vaccines that use various viruses and bacteria as carriers are currently being tested. One concern about carrier vaccines is that the vaccination may need to be repeated (boosted) in order to be effective. Unfortunately, boosting doesn't appear to work with carrier vaccines. This is because vaccination with a carrier virus usually produces enough memory B and T cells to "protect against" a second, booster vaccination. Of course, one way to overcome this problem is to use a different carrier for the booster vaccination. Another is to employ a carrier vaccine for the initial vaccination, and a subunit vaccine to boost the number of memory B cells that are made. This strategy (called "prime-boost") was used in a recent vaccine trial in Thailand.

In this trial, a canarypox virus (a cousin of vaccinia virus) was engineered to produce several HIV-1 proteins, and used as the "priming" vaccine. This vaccination was then boosted by administering a subunit vaccine containing one of the same HIV-1 proteins produced by the carrier virus. Roughly half the people in this trial received the prime-boost vaccinations, whereas the other half received a sham (placebo) vaccination. After three years, the two groups were compared to determine how many in each became infected with the AIDS virus as a result of risky sexual behavior. The authors claimed that the trial "showed a significant, though modest, reduction in the rate of HIV-1 infection." However, many believe that the results are not very convincing. First, only a small number of people (132) became infected during the trial, and the difference between the number of people who received the authentic vaccine and became infected (56) and the number who received the sham vaccine and were infected (76) was small. Moreover, it is suspicious that killer T cells were detected in less than 20% of the people who received the prime-boost vaccine, and there was no significant difference in the amount of virus in the blood of infected members of the two groups. Clearly a larger trial will be required to confirm this result.

POST-INFECTION VACCINES

We tend to think of a vaccination as something one gets for protection against a possible future exposure to a disease-causing agent. However, in some cases, vaccines also can be used to treat a person who has already been infected. The best known example of such a vaccine is the rabies vaccine, which usually is given after a person has been bitten by a rabid animal. This vaccine works well because rabies virus reproduces very slowly. Indeed it may take a month or more before disease symptoms appear. Consequently, the post-infection vaccine can activate the immune system before the rabies virus has a chance to take over.

Another virus for which a post-infection vaccine is now available is varicella-zoster virus (VZV)—a relative of herpes simplex viruses. This virus is named after the two diseases it causes: Varicella is the fancy name for chickenpox, and zoster is what ordinary people call shingles. One important feature which herpes simplex virus and varicella-zoster virus have in common is the ability to establish a latent, stealth infection in their hosts. During the initial infection, VZV infects and replicates in the epithelial cells of the skin. It is this replication and the resulting killing of these cells which causes the rash that is characteristic of chickenpox. In addition, like herpes simplex virus, VZV also is able to infect the sensory nerve cells which are located in the neighborhood of the original skin lesions. And once these nerve cells have been infected, VZV can "hide" there for the life of the host. In contrast to herpes simplex, however, which usually reactivates from time to time to cause repeated episodes of blisters or cold sores, the latent varicella-zoster virus usually causes no further trouble. Nevertheless, in about 15% of people who are over 50 years old and who have been infected with VZV, the virus reactivates and causes shingles.

Immunologists believe that during the life of a person latently infected with VZV, the virus reactivates from time to time and infects some nearby epithelial cells. Usually, the immune system, which remembers the original VZV infection and is "standing guard," destroys the small number of infected cells before the virus can do much damage. However, as we get older, the strength of our immune system declines, and the memories of past infections fade. Consequently, our aging immune memory is less able to deal with reactivation episodes when they occur. This view is consistent with the observation that very few people who are younger than 50 get shingles, and that both the probability of being afflicted with zoster and the severity of the disease increase dramatically with age.

In 1995, a vaccine (Varivax) was licensed to help prevent chickenpox in children, and this vaccine is now included in the battery of vaccines children in the United States routinely receive. Varivax is prepared from an attenuated strain of VZV which is still strong enough to protect about 75% of vaccinated children. Moreover, children who are not completely protected by the vaccine usually experience a milder form of chickenpox than those who have not been vaccinated.

Recently, Varivax has been reformulated to have more than 10 times the potency of the original chickenpox vaccine, and this vaccine (Zostavax) is now offered to adults who are 60 years of age or older. Most of these people were infected with chickenpox as children, and the purpose of this vaccine is to boost their immunity. Studies show that this post-infection vaccine reduces the incidence of shingles by about a factor of two, and significantly decreases the severity of the disease in those who are not completely protected. Because at least one million new cases of shingles occur every year in the United States, a post-infection vaccine which has the potential to prevent even 50% of these cases is quite useful. To put this result in perspective, only about half of the senior citizens who receive the annual flu vaccine are protected by that vaccination—because their immune systems are too weakened by age to respond to the vaccine.

PROSPECTS FOR AN EFFECTIVE AIDS VACCINE

The prospects for developing a vaccine that will protect the general population against an HIV-1 infection are not that great. The virus' ability to establish a latent infection, which is invisible to the immune system, is a serious problem. Recent studies indicate that there is a period of only 5–10 days during which infected cells could be destroyed by the immune system before "unreachable" reservoirs of latently infected cells are established. In addition, HIV-1's high mutation rate makes it an elusive target. Consequently, there is a fear that vaccination against the AIDS virus may not be possible: The human immune system simply may not be capable of mounting a successful defense against this particular virus.

Although HIV-1 poses difficult problems for vaccine development, it certainly is not the only virus against which there is no effective vaccine. Indeed, good vaccines exist for only seven of the 12 viruses in our Bug Parade. For example, herpes simplex virus infects about one-third of the world's population, yet there is no vaccine which can protect against a herpes simplex infection.

GENERAL PRINCIPLES

1. The goal of any vaccination is to generate memory B and/or T cells, for these are the cells that can provide powerful protection if the vaccinated person is subsequently exposed to the "real thing."
2. One important characteristic of vaccines made from "killed" viruses is that, although the chemical treatment will certainly disable most of the viruses, it is impossible to guarantee that it will kill them all.
3. Subunit vaccines have the advantage that, because only one or a few viral proteins are present, there is no possibility that the vaccination will result in an infection.
4. To efficiently produce memory killer T cells, a vaccine must actually be able to infect cells.
5. Vaccines made from attenuated (weakened) microbes can produce memory killer T cells. Moreover, they usually provide lifelong immunity.
6. Although a carrier vaccine infects cells and should produce memory killer T cells, there is no chance that the vaccine will cause the disease – because only a few of the pathogenic virus' genes are "carried" by the vaccine.

THOUGHT QUESTIONS

1. Review the various types of vaccines and discuss whether each type might be used to protect the general public against the "Andromeda Strain."

Chapter 19

Antiviral Drugs

BACKGROUND

One reason to try to understand how viruses "think" is that this information might help virologists out-think them. For example, given sufficient information, it might be possible to intervene to block access to, or entry into, the virus' target cells; to disrupt viral reproduction; to counteract the tactics the virus uses to evade host defenses; or to prevent the spread of the virus to new hosts.

Although vaccines have been extremely useful in protecting against viral attacks, once a person has been infected, the drugs available to treat the infection are limited. At first glance, this doesn't seem to make much sense. After all, antibiotics are used to treat a wide range of bacterial infections. So why are there no broad-spectrum antiviral drugs with that same capability?

Bacteria tend to be "free-living" organisms, which have evolved ways of doing business that are sometimes very different from the ways things are done in human cells. For example, the walls of human cells and bacterial cells are assembled from very different materials. As a result, an antibiotic such as penicillin, which disrupts the synthesis of a common bacterial cell wall component, can kill a wide range of bacteria without damaging human cells.

In contrast, viruses are dependent on the biochemical machinery of human cells for their reproduction. Consequently, it is very difficult to discover drugs that will kill viruses but not human cells. Indeed, most antiviral drugs have significant side effects that make them unsuitable for long-term use in chronic viral infections. Also, because individual viruses have solved their common problems in so many different ways, it is difficult to imagine the possibility of a broad-spectrum antiviral drug. Indeed, those lifestyle features that are uniquely viral generally are common to only a small number of viruses.

The lack of broad-spectrum antivirals is a real problem. Many bacterial infections cause similar symptoms, but by treating these infections with broad-spectrum antibiotics, the exact identity of the infecting bacterium frequently need not be determined. In contrast, the lack of broad-spectrum antivirals usually means that the invading virus must be identified before treatment can begin. And in many cases, this identification takes so long that by the time the appropriate antiviral drug can be selected, an acute viral infection will already have run its course—or a chronic or latent infection will already have been established.

Broad-spectrum antivirals aside, it is difficult even to produce an antiviral drug that is effective against a single virus. To rationally design an antiviral drug, virologists must first identify a target which is uniquely viral. This usually requires a detailed understanding of the virus' lifestyle, and this information is only available for a small number of viruses. Consequently, many of the current treatments were discovered by testing "every drug on the shelf" for antiviral activity.

As a first step in determining the efficacy of a potential antiviral compound, virologists would like to test virus-infected human cells in the lab to determine whether treatment with the antiviral reduces virus production while sparing uninfected cells. However, for several of the most important viruses (e.g., hepatitis B and the human papillomavirus), good systems do not exist for growing the virus in the lab.

If a compound works well on virus-infected cells, virologists would next like to test whether the antiviral will be effective in an animal. Such experiments are vital to determine whether the drug will reach the right part of the body to work its magic, and whether it can be maintained in the animal at an effective concentration without being toxic. Unfortunately, good animal models for viral infection do not exist for some very important viruses (e.g., HIV-1).

Finally, drug development is insanely expensive and time consuming: It usually costs at least 100 million dollars and takes 5 to 10 years to test and license a new antiviral drug for general use. This fact is often overlooked by consumers who complain that prescription drugs cost too much.

TARGETS FOR ANTIVIRAL DRUGS

The goal of an antiviral drug is to interfere with some step during a viral infection that is uniquely viral. These targets can be arranged into four main groups, corresponding to the problems every virus must solve: entry and uncoating, viral reproduction, viral evasion of host defenses, and spread to new human hosts. In our discussion here, we will evaluate how successful virologists have been in developing drugs that target each of these stages.

Interfering With Viral Entry and Uncoating

Neutralizing antibodies are able to prevent entry of some viruses into their target cells. Consequently, it would seem that drugs could be discovered or designed which could block viral entry. Part of the problem with this approach is that some viruses (e.g., herpes simplex and HIV-1) have alternative cellular receptors which they can use to gain entry. So a combination of drugs would be needed which could block the interaction of a virus with all of its possible receptors.

Another difficulty in designing therapeutic drugs which prevent viral entry is that it involves a "numbers game" which is heavily weighted against the drug. A viral capsid or envelope contains many copies of the proteins which bind to their cellular receptors, and the surface of a cell usually has thousands of these receptors. To block entry, a drug would have to interact with a significant fraction of the viral proteins or cellular receptors—and that's a tall order. In some cases, antibodies can win this numbers game because a huge number of antibody molecules can be produced by B cells. Moreover, antibody molecules are extremely stable, with half lives that usually are measured in days or weeks. In contrast, high concentrations of most therapeutic drugs are toxic, and many drugs persist within the body for only a few hours before they are inactivated (e.g., by binding to proteins in the blood), excreted in the urine, or degraded in the liver.

Despite these obstacles, a drug which can block the entry of HIV-1 into its target cells has now been approved by the FDA. The drug, MVC (maraviroc), is a small, synthetic molecule which binds to the CCR5 co-receptor which HIV-1 frequently uses for entry. This binding distorts the conformation of CCR5 so that it no longer interacts with the gp120 protein on the surface of the virus. Nevertheless, this drug is not completely effective, because HIV-1 can use another co-receptor, CXCR4, to gain entry into at least some of the cells it infects.

After a virus binds to its target cell, it must shed its coat either during or after entry. An excellent example of a therapeutic drug which interferes with viral uncoating is amantadine. This drug, which can be used to treat influenza A infections, is now over 30 years old. It inhibits the activity of a viral protein, M2, which is required for uncoating. The envelope of influenza A virus contains a relatively small number of M2 proteins (fewer than 70). Moreover, multiple M2 proteins must fit together properly to form a channel through which protons can enter the interior of the virus to facilitate uncoating. As a result, relatively few M2 molecules need to be "compromised" by the drug to keep the ion channels from working properly and to trap the virus within its coat.

Unfortunately, even this highly specific antiviral has its problems. First, the influenza virus mutation rate is so high that essentially every virus produced is different from the original infecting virus. As a result of this antigenic drift, influenza mutants arise whose M2 proteins no longer interact with amantadine. Second, to be effective in decreasing the severity of an influenza attack, amantadine must be administered during the first two days after infection—and most people don't even realize they have the flu for the first day or two. Moreover, because other microbes cause flu-like symptoms, it is difficult for a physician to be sure that amantadine is the drug of choice. For example, influenza B virus causes early symptoms similar to those caused by influenza A virus, yet influenza B virus, which has no M2 protein, is unaffected by amantadine. Consequently, amantadine is usually prescribed to protect against an anticipated influenza A infection rather than as a treatment.

Interfering With Viral Reproduction

Although viruses rely on cellular machinery for their reproduction, all viruses replicate their genomes using strategies which are at least somewhat different from that used for the replication of cellular DNA. These idiosyncrasies in viral genome replication represent excellent targets for antiviral therapies.

AZT is a drug that takes advantage of a unique feature of viral replication, and which can be used to treat HIV-1 infections. When HIV-1 replicates, it employs its reverse transcriptase enzyme to make a complementary DNA copy of its RNA genome. During this operation, DNA building blocks (the nucleotides) are strung together according to the genetic code contained within the single strand of HIV-1 RNA. Each nucleotide has a "plug" (the 5′ tri-phosphate) and a "socket" (the 3′ hydroxyl), and new nucleotides are added by plugging them into the socket of the nucleotide that is at the end of the growing DNA chain. AZT, a nucleoside analog, is converted by cellular enzymes into a "fake nucleotide"—one which has a plug, but no socket. Consequently, when AZT is incorporated into DNA, there is no socket for the next nucleotide to plug into, and the DNA chain terminates. The result of this premature termination is a viral cDNA which cannot be integrated into the cellular DNA of the infected cell—and without integration, the viral infection is aborted. Because it interferes with reverse transcription of the HIV-1 genome, AZT is designated as a "nucleoside reverse transcriptase inhibitor."

Unfortunately, although AZT can interfere with HIV-1 replication, it also can be incorporated into cellular DNA in proliferating human cells. When this happens, these cells can be killed. Fortunately, the HIV-1 reverse transcriptase has a higher affinity for the AZT nucleotide than do most cellular polymerases. Consequently, there is a "therapeutic window" of AZT concentrations which favors incorporation into viral versus human DNA. Nevertheless, there are many cells in the body which proliferate almost continuously (e.g., cells that line the digestive tract and blood cells in the bone marrow), and these cells can be killed by the AZT treatment, producing substantial side effects. This problem is exacerbated by the fact that the half life of AZT in a human is only about one hour, because the drug is rapidly degraded by the liver. So keeping the AZT concentration high enough to interfere with viral replication without producing serious side effects is a challenge. In addition, because it is difficult to maintain a therapeutic level of AZT, the virus continues to replicate when concentrations wane—and mutant viruses arise whose reverse transcriptase enzymes will no longer incorporate AZT into their cDNAs. For these reasons, AZT used alone is only moderately useful in treating a chronic HIV-1 infection. In contrast, a short course of AZT treatment is usually quite effective in reducing the probability that an infected mother will transmit the AIDS virus to her child at birth.

Acyclovir is another nucleoside analog that can act as a DNA chain terminator. The real beauty of acyclovir, however, is that the conversion from nucleoside analog to fake nucleotide is not carried out efficiently by cellular enzymes, so uninfected cells are rarely damaged by this drug. In contrast, both herpes simplex virus and varicella-zoster virus encode an enzyme, thymidine kinase, which, together with cellular enzymes, efficiently converts acyclovir into a fake nucleotide. As a result of this selectivity, acyclovir is very useful for treating herpes simplex and varicella-zoster infections.

Ribavirin is a nucleoside analog which does not function by terminating nucleic acid synthesis. This drug has both a plug and a socket. Indeed, until very recently, the mode of action of this antiviral drug was quite mysterious. Now it has been discovered that ribavirin acts by increasing the rate at which viral RNA polymerase enzymes introduce mutations into the RNA molecules they make. Here's how this works.

RNA viruses have a high mutation rate because of the error-prone nature of their RNA polymerases. This high rate of mutation is a great benefit to these viruses, because it makes it more likely that the viral population created during an infection will include mutants which can resist newly evolving host defenses. However, this high mutation rate can also be dangerous to the virus: If the rate is too high, many newly made viruses will be so badly mutated that they cannot reproduce. So RNA viruses must walk a fine line between mutating too slowly and falling prey to evolving host defenses, and mutating so rapidly that they become nonfunctional. It has been proposed that the best solution to this problem is for a virus to "adjust" its mutation rate to be slightly below the point at which the errors introduced begin to seriously decrease the viability of the virus. What this means is that if one could treat virus-infected cells with a drug which would substantially increase the already high viral mutation rate, the RNA viruses replicating in those cells might be forced into "error catastrophe." And that's just what ribavirin does.

When ribavirin is taken up by a cell, it is phosphorylated by host enzymes to produce a fake G (guanosine) RNA nucleotide. Once ribavirin has been incorporated into the position normally occupied by a real G nucleotide, it can be copied along with the rest of the RNA molecule when the virus replicates. However, whereas the real G nucleotide templates the addition of a C (cytidine) nucleotide to the growing, complementary RNA chain, ribavirin templates the addition of C or U (uridine) nucleotides with roughly equal efficiencies. And introducing a U instead of a C results in a mutation. If the concentration of ribavirin within the cell is high enough, the virus may be driven into an error catastrophe in which most of the newly made RNA genomes are non-functional. Reaching the concentration required for lethal mutagenesis is made easier by the fact that ribavirin also inhibits a cellular enzyme required to produce real G nucleotides. This decreases the pool of real G nucleotides within the cell, and increases the probability that the fake G will be incorporated into growing viral genomes.

Although ribavirin has been used to treat a number of different viral infections and is frequently cited as a broad-spectrum antiviral, it has only been conclusively demonstrated to be efficacious in treating hepatitis C infections (in conjunction with interferon) and respiratory syncytial virus infections in infants. It is hypothesized that ribavirin's lack of clear broad-spectrum activity may result from the differing efficiencies with which various viral RNA polymerases incorporate the fake bases into their genomes. Alternatively, different viruses may be more or less susceptible to the mutagenic activity of ribavirin because they have chosen to "walk" at different distances from the error catastrophe "precipice."

In addition to drugs like AZT and acyclovir, which are incorporated into growing viral DNA chains, there are also drugs which interfere with synthesis of viral DNA by binding to the viral polymerase molecule. One such drug, foscarnet, binds to the polymerase enzymes of hepatitis B virus, HIV-1, and certain herpes viruses (e.g., herpes simplex, human cytomegalovirus, vericella-zoster virus). Even though the polymerases of these viruses are very different, this drug can bind to them all, so foscarnet is about as close to a broad-spectrum antiviral as has been discovered. However, foscarnet is a highly charged molecule, so transport across the cell membrane is inefficient. As a result, therapeutic doses usually are so high, and the side effects so serious (kidney toxicity) that forscarnet is a "last resort" drug which mainly is used in immediate, life-threatening illnesses.

Another non-nucleoside polymerase inhibitor is nevirapine. This drug binds to the HIV-1 reverse transcriptase at a site just next to the active site of the enzyme, creating a distortion which interferes with the polymerase's ability to produce viral cDNA. There is now a whole family of non-nucleoside, reverse transcriptase inhibitors like nevirapine, and these drugs are used extensively in treating HIV-1 infections.

During the final stages of assembly of HIV-1 virions, a viral enzyme (the HIV protease) cuts a large Gag-Pol precursor protein into pieces to yield viral structural proteins and enzymes. This protease activity is unique to the virus (i.e., there is no cellular enzyme that can do this cutting), so it has become an attractive target for anti-HIV-1 drugs. One representative protease inhibitor, indinavir, mimics a cleavage site that the HIV protease recognizes in the Gag-Pol protein. At sufficiently high concentrations, indinavir can "distract" the viral protease and keep it from making the cut in the Gag-Pol protein required for viral assembly.

Hepatitis C virus encodes a protease (NS3-4A) which is required to cleave the viral polyprotein into smaller proteins. Recently, the FDA approved two drugs, boceprevir and telaprevir, which block the action of this protease. Clinical trials suggest that either of these protease inhibitors, when used together with the current standard drug therapy (ribavirin plus IFN-α), can cure roughly 75% of hepatitis C infections.

During an HIV-1 infection, three viral enzymes are produced which have no counterpart in human cells, and which, therefore, are excellent targets for chemotherapy. We have discussed drugs which target two of these enzymes: the reverse transcriptase and the protease. The third enzyme is the integrase protein, which is essential for preparing the viral cDNA for insertion into the cell's chromosome. There is now a drug, raltegravir, that interferes with the action of the HIV-1 integrase, and which is being used to treat AIDS patients.

Overcoming Viral Evasion of Host Defenses

Type 1 interferons are "natural" inhibitors of viral reproduction, and viruses have evolved many clever ways to protect themselves against the interferon defense. Consequently, there are only a few viruses for which interferon treatments have been successful. Interferon-α has been used to treat hepatitis B and hepatitis C infections, but less than 50% of those treated showed a long-term response to this drug.

Virus Exit Inhibitors

In order to spread to new hosts, a virus must exit the cell in which it is produced—a process that is not as easy as it might seem. For example, when influenza virus exits a cell it has infected, there is a problem it must solve. Because the viral envelope contains both hemagglutinin and hemagglutinin receptors, there is a danger that influenza viruses will bind to each other as they exit, producing non-infectious clumps of virus. In addition, the hemagglutinin proteins on the exiting viruses can bind to receptors on the cell they are leaving, trapping the virus on the surface of a cell that has already been infected. Influenza virus deals with this exit problem by producing a razor-like protein, the viral neuraminidase. This protein is inserted into the cell membrane, where it "shaves" the cellular receptors, removing the sialic acid residues to which the hemagglutinin proteins bind. Because this neuraminidase activity is so important for viral spread, the viral neuraminidase protein is an excellent target for anti-influenza drugs.

Zanamivir (Relenza) is a drug which is inhaled, and oseltamivir (Tamiflu), is taken orally. Both are neuraminidase inhibitors, and both are effective against both type A and type B influenza. These drugs are sialic acid mimics to which the neuraminidase enzyme binds much more tightly than to real sialic acid residues. As a result, the neuraminidase enzyme tries to shave the drug, rather than the cell. If these drugs are used within the first two days after infection, they can shorten the duration of symptoms by a day or two. Of course, as we discussed earlier, it is difficult to self-diagnose an influenza infection within the first 48 hours, so these drugs generally are used for people living in close proximity after one or more of them has already been diagnosed with an influenza infection (e.g., during a flu outbreak in a nursing home). In such cases, the drug is about 60% effective in preventing infection.

Combination Therapies

During replication, many viruses mutate rapidly, and these mutations can render antiviral drugs ineffective. Indeed, drug-resistant mutants have been described for every antiviral drug in common use today. It is important to remember that viral mutants arise only when viruses replicate. This means that if a drug does not stop viral replication completely, a virus will be free to mutate so as to "escape" the effects of the drug. Escape mutants are not such a big problem in acute infections (e.g., an influenza infection), because such infections are usually dealt with so quickly by host defenses that the probability of an escape mutant arising is relatively small. In contrast, with chronic infections, there is usually enough time (i.e., enough viral replication cycles) to insure that an escape mutant will eventually be produced. For example, as early as 1989, it was observed that AIDS patients who had been treated with AZT for 6 months or longer frequently harbored escape mutants which were resistant to this drug. The realization that a single drug would probably not be effective in controlling an HIV-1 infection led to the suggestion that multiple drugs should be used simultaneously to treat AIDS patients. The hypothesis was that if a patient were treated with several drugs at the same time, it would be less likely that a single viral genome would mutate in such a way as to simultaneously confer resistance to all the drugs. Indeed, certain combinations of drugs have been found which can greatly decrease HIV-1 replication, and which can extend the life of an infected individual.

With the discovery of non-nucleoside reverse transcriptase inhibitors and HIV protease inhibitors, it was natural to try to combine these with nucleoside reverse transcriptase inhibitors in a drug "cocktail." Treatment with such a cocktail of anti-HIV-1 drugs is referred to as highly active antiretroviral therapy (HAART). If one had to guess, one might predict that the best combination of three drugs to use in HAART might be one nucleoside reverse transcriptase inhibitor, one non-nucleoside reverse transcriptase inhibitor, and one protease inhibitor. However, so far it has turned out that the best combinations are two nucleoside reverse transcriptase inhibitors plus either one non-nucleoside reverse transcriptase inhibitor, one protease inhibitor, or one integrase inhibitor. A drug cocktail, Atripla (efavirenz, emtricitabine, and tenofovir), which includes two nucleoside reverse transcriptase inhibitors, and one non-nucleoside reverse transcriptase inhibitor, is available in pill form to be taken only once a day.

Although HAART can extend the life of many AIDS patients and can delay the onset of opportunistic infections, it is not without its problems. First, HAART is not a cure. HIV-1 can persist for very long periods with its provirus integrated into the genomes of infected cells. Because the drugs used in HAART target virus replication, non-replicating proviruses cannot be touched by any of the available drugs. Indeed, latently infected cells represent a reservoir from which the virus can "bounce back" once HAART is discontinued. Also, none of the drugs used for HAART is without considerable side effects, and there is a tendency for patients to discontinue HAART or to adhere less strictly to the treatment schedule once they begin to feel better—and non-adherence can lead to the resumption of viral replication and eventually to drug resistance. Further, these drugs are not cheap. A year of HAART typically costs about $20,000, putting this treatment beyond the reach of many patients.

GENERAL PRINCIPLES

1. Viruses are dependent on the biochemical machinery of human cells for their reproduction. Consequently, it is very difficult to discover drugs that will kill viruses, but not human cells.
2. Because individual viruses have solved their common problems in so many different ways, it is difficult to imagine the possibility of a broad-spectrum antiviral drug.

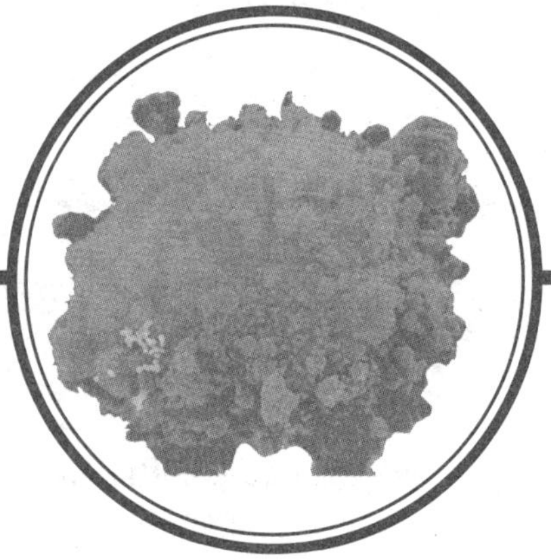

Summary Tables

Table I Viruses We Inhale

	Influenza A "Bait-and-Switch" Virus *Orthomyxoviridae* Family	**Rhinovirus** A Virus That Surrenders *Picornaviridae* Family	**Measles** A "Trojan Horse" Virus *Paramyxoviridae* Family
INFECTION	Infects epithelial cells in upper and lower respiratory tract	Mainly infects epithelial cells in upper respiratory tract	Infects dendritic cells in the airway
	Infects in overwhelming numbers to breach mucosal barrier	Infects in overwhelming numbers to breach mucosal barrier	Tricks dendritic cells into carrying virus into lymph nodes
	Uses mucosal enzymes to prepare for entry	Uses mucociliary escalator to reach target cells	Infects additional immune system cells in lymph nodes and establishes a systemic infection
	Entry via receptor-mediated endocytosis	Binds to target cell and releases genome directly into cytoplasm	Binds to cellular receptor, its envelope fuses with the cell membrane, and its genome is released into cytoplasm

(continues)

Table I Viruses We Inhale (*continued*)

	Influenza A "Bait-and-Switch" Virus *Orthomyxoviridae* Family	Rhinovirus A Virus That Surrenders *Picornaviridae* Family	Measles A "Trojan Horse" Virus *Paramyxoviridae* Family
REPRODUCTION	Negative, single-strand, segmented RNA genome; replicated in nucleus	Positive, single-strand RNA genome; not segmented; translated into one long polyprotein	Negative, single-strand RNA genome; not segmented
	Error-prone RNA polymerase	Error-prone RNA polymerase	Viral polymerase is error-prone, but mutations in hemagglutinin cannot be tolerated
	Steals caps to favor synthesis of viral proteins	Uses internal ribosome entry site for cap-independent translation of viral mRNA to favor synthesis of viral proteins	Viral polymerase stops and restarts to synthesize mRNAs
	Picks up envelope, and buds from surface	Has single capsid made of protein	Picks up envelope, and buds from surface
	Cytolytic	Cytolytic	Cytolytic
EVASION	Protein NS1 blocks recognition of viral RNA by pattern recognition receptors	Rapidly shuts down host protein synthesis before much interferon can be produced	Blocks interferon production by interfering with IRF3 phosphorylation
	NS1 blocks the action of interferon-induced, antiviral proteins	Blocks transport of interferon to the cell surface	Diminishes effects of interferon by disrupting signal from interferon receptor to nucleus of infected cell
	Reproduces rapidly to outrun immune system during initial infection	Reproduces quickly and then surrenders to innate defenses before the adaptive system is fully activated	Immunosuppresses host, and spreads by cell fusion to evade adaptive immune defense
SPREAD	Inflammation caused by immune system generates both cough and sneeze to spread virus	Infection of upper respiratory tract triggers sneezes, facilitating spread	Spreads when systemic infection returns virus to respiratory tract, and triggers sneeze and cough reflexes
	Bait and switch — uses antigenic drift and antigenic shift to make return visits possible	Weak adaptive system response and antigenic drift allow virus to return to infect same host	Only one serotype, so only infects each human once; can only be maintained in a large population
	Causes acute infection	Causes acute infection	Causes acute infection

Table II Viruses We Eat

	Rotavirus An Undercover Virus *Reoviridae* Family	**Enteric Adenovirus** Virus With a Time Schedule *Adenoviridae* Family	**Hepatitis A** A Virus That Detours *Picornaviridae* Family
INFECTION	Infects cells at tips of intestinal villi	Infects intestinal epithelial cells	Initiates infection by gently infecting intestinal cells
	Uses digestive enzymes to prepare for entry into cell		Carried by IgA antibodies to liver where it infects hepatocytes
	Binds to receptor, is enclosed by an endosome, and taken into cell	Binds to receptor and co-receptor, is enclosed in an endosome, and is taken into cell	Binds to target cell and releases genome directly into cytoplasm
REPRODUCTION	Segmented, double-stranded RNA genome	Large, linear, double-stranded DNA genome; viral DNA polymerase replicates both strands continuously	Positive, single-stranded RNA genome; not segmented
	Genome enclosed in three protein shells (capsids)	Single protein capsid	Single protein capsid
	mRNA transcribed "under cover" while genome is still inside two capsids	Reproduces slowly, using carefully timed plan of viral gene expression	mRNA is translated to produce one long polyprotein
	NSP3 interferes with translation of polyadenylated cellular mRNA	Takes over control of DNA and protein synthesis in infected cells	Uses internal ribosome entry site for cap-independent translation of viral mRNA
	Cytolytic	Cytolytic	Non-cytolytic

(continues)

Table II Viruses We Eat (*continued*)

	Rotavirus An Undercover Virus *Reoviridae* Family	Enteric Adenovirus Virus With a Time Schedule *Adenoviridae* Family	Hepatitis A A Virus That Detours *Picornaviridae* Family
EVASION	Replicates undercover to avoid activation of interferon system	Defends against effects of interferon by making VA "decoy proteins" which compete with viral double-stranded RNA for binding to PKR	Viral proteins suppress interferon production by blocking TLR3 and RIG-I pathways.
	NSP1 degrades transcription factors required for interferon production	Evades adaptive immune system by interfering with viral antigen presentation by MHC molecules	Confuses the adaptive immune system by "detouring" through the liver
	Reproduces quickly before the adaptive immune system can be fully mobilized	Viral proteins hold off apoptosis until replication is complete	
SPREAD	NSP4 induces diarrhea to transport newly made viruses to outside world; virus remains infectious in water or on dry surfaces	"Death protein" explodes infected cells; diarrhea transports newly made viruses to outside world	Virus produced in liver is carried by bile to the intestine to exit with feces; no diarrhea
	Error-prone polymerase creates antigenic drift; segmented genome and animal hosts makes antigenic shift possible	Error-prone polymerase creates antigenic drift	Only one serotype — depends on lax hygiene to spread; survives well in dried feces
	Causes acute infection	Causes acute infection	Causes acute infection

Table III Viruses We Get From Mom

	Hepatitis B Virus A Decoy Virus *Hepadnaviridae* Family	Hepatitis C Virus An Escape Artist *Flaviviridae* Family	HTLV-I A Tribal Virus *Retroviridae* Family
INFECTION	Extremely efficient infection by blood-to-blood contact	Infection by blood-to-blood contact	Infection through transfer of infected cells by breastfeeding, sex, and contaminated blood
	Infects liver cells (hepatocytes)	Infects liver cells (hepatocytes); entry requires at least four cellular receptor proteins	Infects immune system cells, including helper T cells and dendritic cells; entry requires three cellular receptor proteins
	Viral envelope binds unidentified receptor, fuses with cell plasma membrane, and encapsidated genome is released into cytoplasm	Viral envelope fuses with cell plasma membrane, and encapsidated genome is released into cytoplasm	Viral envelope fuses with cell plasma membrane, and two copies of genome are released into cytoplasm
	Humans are the only host	Only hosts are humans and chimpanzees	Humans are the only host
REPRODUCTION	Circular, mostly double-stranded DNA genome	Positive, single-stranded RNA genome; not segmented; translated into one long polyprotein	Single-stranded RNA genome is reverse-transcribed by viral enzyme to make a double-stranded DNA provirus which is integrated into chromosome of cell
	Bizarre replication strategy using viral reverse transcriptase enzyme	Uses internal ribosome entry site for cap-independent translation	Uses cellular RNA polymerase to transcribe provirus to make viral mRNA and viral genomes
	Has protein capsid enclosed in a lipid envelope derived from endoplasmic reticulum	Has protein capsid enclosed in a lipid envelope derived from endoplasmic reticulum	Has protein capsid enclosed in a lipid envelope derived from cell's plasma membrane
	Non-cytolytic	Non-cytolytic	Non-cytolytic
EVASION	Efficiently blocks interferon production	Blocks interferon production, interferes with interferon stimulated gene expression, and blocks function of interferon-induced, antiviral proteins	Provirus integrated into cellular DNA can establish latent infection with little or no viral protein synthesis
	Produces empty "decoy" viruses to confuse antibody defense	Error-prone polymerase helps virus escape from the adaptive immune system	To evade killer T cells, viral protein (p12) interferes with display of viral proteins by class I MHC molecules

(continues)

Table III Viruses We Get From Mom (*continued*)

	Hepatitis B Virus A Decoy Virus *Hepadnaviridae* Family	Hepatitis C Virus An Escape Artist *Flaviviridae* Family	HTLV-I A Tribal Virus *Retroviridae* Family
SPREAD	Spreads within host by infection with newly made virions	Spreads within host by infection with newly made virions and by cell-to-cell contact	Spreads within host mainly by proliferation of cells with integrated provirus or by cell-to-cell transmission
	Transmitted by infected mother to her child during birth or from child to child during rough play	Transmitted by infected mother to her child during birth or from child to child during rough play	Transmitted by infected mother to her child during breastfeeding or by infected adult during sex
	Transmitted by reused needles or by contaminated blood transfusions	Transmitted by reused needles or by contaminated blood transfusions	Transmitted by reused needles or by contaminated blood transfusions
	Causes either an acute infection or a smoldering chronic infection of the liver	Causes either an acute infection or a smoldering chronic infection of the liver	Causes a systemic, chronic infection with many latently infected cells

Table IV Viruses We Get By Intimate Physical Contact

	HIV-1 An Urban Virus *Retroviridae* Family	Herpes Simplex Virus A Virus That Hides *Herpesviridae* Family	Human Papillomavirus A Very Quiet Virus *Papillomaviridae* Family
INFECTION	Infects $CD4^+$ cells (helper T cells, dendritic cells, macrophages) in vagina or anus, travels to lymph nodes, and establishes systemic infection	Replicates efficiently in epithelial cells; infects, but produces little or no virus in nerve cells	Infects basal epithelial cells, but produces no virus until these cells "mature" and move toward surface of the epithelium
	Virus binds to CD4 protein and to a co-receptor on cell surface	Virus binds heparan sulfate on cell surface; localized infection — not systemic	Localized infection — not systemic
	Viral envelope fuses with cell plasma membrane, and encapsidated genome is released into cytoplasm	Viral envelope fuses with cell plasma membrane, and encapsidated genome is released into cytoplasm	Virus taken into cell by receptor-mediated endocytosis
	Humans and chimpanzees are only hosts	Humans are only hosts	Humans, of course
REPRODUCTION	Single-stranded RNA genome is reverse-transcribed by viral enzyme to make a double-stranded DNA provirus which is integrated into chromosome of cell	Large, linear, double-stranded DNA genome is replicated as a "rolling circle" — even in resting epithelial cells	Small, circular, double-stranded DNA genome; replicates as episome within cell's nucleus
	Uses cellular RNA polymerase to transcribe integrated provirus to make viral mRNA and viral genomes	Uses cellular RNA polymerase to transcribe viral mRNA; inhibits splicing of cellular mRNA	Uses cellular RNA polymerase to transcribe viral mRNA
	Has protein capsid enclosed in a lipid envelope derived from cell's plasma membrane	Has capsid, tegument, and cell-derived envelope	Has single capsid made of protein
	Cytolytic	Non-cytolytic in nerve cells; cytolytic in epithelial cells	Non-cytolytic

(continues)

Table IV Viruses We Get By Intimate Physical Contact (*continued*)

	HIV-1 An Urban Virus *Retroviridae* Family	Herpes Simplex Virus A Virus That Hides *Herpesviridae* Family	Human Papillomavirus A Very Quiet Virus *Papillomaviridae* Family
EVASION	Provirus integrated into cellular DNA can establish undetectable, latent infection in $CD4^+$ cells	Establishes undetectable, latent infection of nerve cells	Establishes undetectable, latent infection of basal epithelial cells
	Virus particles can be transferred from infected cell to uninfected cell, avoiding antibody defense	Viral proteins interfere with action of complement proteins and antibodies	Reproduction is non-cytolytic; usually evades adaptive immune system
	Induces interferon production to activate immune system cells, making them better targets for infection	Viral proteins protect against effects of interferon system	To avoid detection by interferon system, mRNA is transcribed from only one DNA strand
	High mutation rate allows virus to outrun adaptive immune system; turns immune system against itself; dysregulates and destroys immune system	Evades killer T cells by interfering with presentation of viral proteins by class I MHC molecules	Viral proteins drive "unscheduled" cell proliferation and protect against the consequences (apoptosis)
SPREAD	Spreads within individual mainly when virus infects $CD4^+$ cells or by transinfection	Usually does not spread within host to sites other than that of initial infection	Passed down to daughter cells when basal epithelial stem cells proliferate
	Efficiency of sexual transmission increases dramatically in context of other sexually transmitted diseases	Spread efficiently by intimate physical contact — vaginal or oral sex	Spread by intimate physical contact when damage to epithelial layer exposes basal cells
	Transmitted by transfer of infected cells or virus	Transmitted by infectious virus	Transmitted by infectious virus
	Establishes a reservoir of latently infected $CD4^+$ cells from which virus can be reactivated	Establishes a reservoir of latently infected sensory nerve cells from which virus can be reactivated	Establishes reservoir of latently infected basal epithelial stem cells
	Causes a systemic, lifelong, chronic infection in which $CD4^+$ cells are continuously killed and replaced	Infection usually lasts a lifetime; infectious virus is produced even when infection is asymptomatic	Infection typically lasts months or a few years

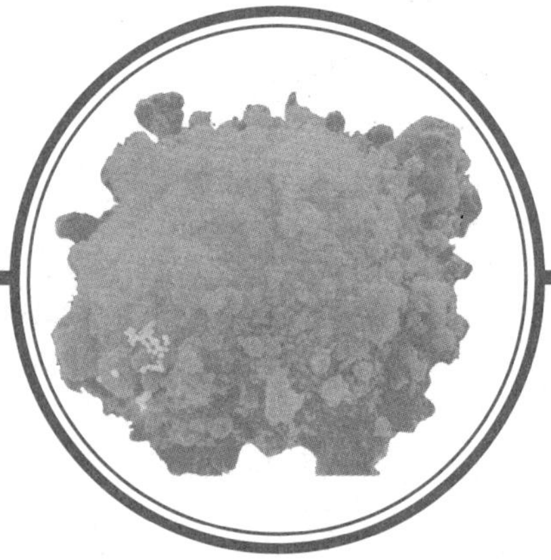

Glossary

A

Abortive infection An infection in which some virus proteins are made, but no infectious virus is produced.

Acute infection An infection in which the virus is banished after a short stay in the host.

Adaptive immune system The part of the immune system that includes B cells (which produce antibodies) and killer T cells (which destroy infected cells).

Antigenic drift Small genetic changes due to errors made in copying a viral genome.

Antigenic shift Drastic genetic changes introduced when gene segments derived from humans, birds, and animals are mixed and matched.

Apoptosis The process by which cells commit suicide when they sense a viral infection or other disruption of normal cellular processes.

C

Capsid A shell-like structure, composed of many copies of a few viral proteins.

Chronic infection A long-term infection in which new viruses are produced, and the host defenses battle to keep the infection in check.

Cytolytic virus A virus that kills the cells it infects.

E

Endosome A "pouch" or vesicle made from a cell membrane.

Enteric virus A virus which infects the digestive tract.

Envelope A lipid bilayer derived from a cell membrane.

Epithelial cells Cells that line the surfaces of the body, including the skin, the respiratory tract, the digestive tract, and the reproductive tract.

G

Genome The sum total of all the genetic information contained in an organism. For example, a virus' genome is the collection of all the virus' genes.

H

Host range The type of cell or species a virus infects.

I

Innate immune system The part of the immune system which includes the interferon system, the complement system, and macrophages.

K

Kinase An enzyme which sticks a phosphate onto another molecule.

L

Latent infection An infection in which the virus lies dormant.

O

Opsonized "Decorated" with fragments of complement proteins or antibodies.

P

Pattern recognition receptors Cellular proteins which detect molecular patterns that are characteristic of viruses and other invaders.

Productive infection An infection in which new viruses are produced.

Proliferation The process in which cells grow to twice their original size, duplicate their genomes, and divide into two daughter cells.

Protease An enzyme that cleaves proteins.

Provirus A double-stranded DNA copy of a retroviral genome which can be pasted into one of the chromosomes of an infected cell.

R

Receptor-mediated endocytosis The process by which a virus binds to a receptor molecule on the cell surface, and is taken into the cell enclosed in a patch of the cell's plasma membrane.

Resting cells Cells that are not proliferating.

RIG-I A pattern recognition receptor which detects viral RNA that is terminated with a 5′ triphosphate.

S

Secrete To export out of a cell into the surroundings.

T

Tegument A layer of proteins in between a herpes virus' capsid and its envelope.

Toll-like receptors Cellular proteins which detect molecular patterns that are characteristic of viruses and other invaders.

V

Virion Synonymous with virus.

Virulence How effective a virus is in causing disease—synonymous with pathogenicity.

Virus particle Synonymous with virus.

Z

Zoonosis A human disease that is the result of an infection by a virus whose usual host is an animal.

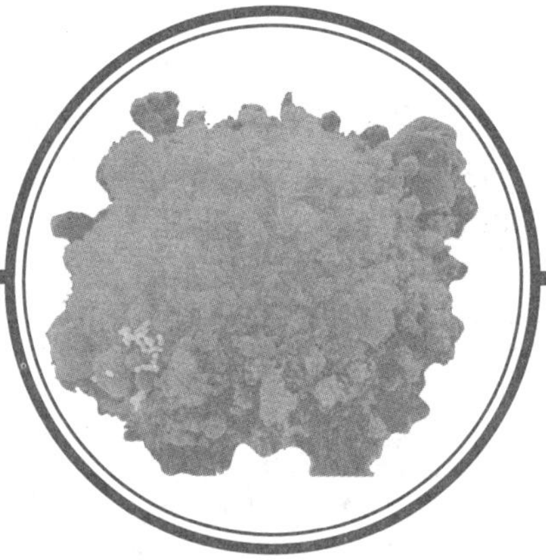

Index

Note: Italicized page locators indicate figures; tables are noted with *t*.

B

C

U

V

W

X

Z